ASTROPHYSICS

ASTROPHYSICS

William K. Rose   *University of Maryland*

Holt, Rinehart and Winston, Inc.
New York Chicago San Francisco Atlanta Dallas Montreal Toronto London Sydney

PREFACE

Progress in astronomy and astrophysics has been rapid in recent years primarily because developments in physics and technology have enabled astronomers to make measurements with far greater precision and sophistication than previously possible, and also because whole new fields of astronomy (for example, infrared and x-ray astronomy) have been initiated. Rapid growth in astronomical knowledge has led to a greatly improved understanding of physical conditions throughout the universe, and consequently far-reaching advances in theoretical astrophysics have ensued.

The astrophysicist attempts to explain astronomical phenomena by means of plausible theories that are consistent with empirical evidence and the laws of physics. The purpose of this book is to describe various topics in astrophysics in such a manner as to provide a coherent introduction to the subject. Much of astrophysics is concerned with the study of self-gravitating bodies. Since stars are the simplest and most conspicuous of such objects, stellar theory has played a unique role in the development of astrophysics. For this reason, a significant fraction of the book is devoted to the study of the birth, evolution, and final states of stars.

Chapter 1 presents a short introduction to cosmology and a description of the major constituents of space. Much of this chapter is concerned with the physical properties of the interstellar medium. The basic properties of stars are discussed in Chapter 2. This chapter includes a discussion of the equations and physical processes that are most important in understanding stellar interiors. Chapters 1 and 2 serve as an introduction to Chapters 3 to 11, which trace the history of stars from their origin to final states. The remarkable discovery of pulsars indicates that neutron stars are the remnants of many supernova explosions. Because of the great interest in pulsars, three chapters (9, 10, and 11) are devoted to supernovae, pulsars, and neutron stars.

Galaxies are massive aggregates of stars. Prior to the last two decades, astrophysics was concerned almost exclusively with the study of objects whose radiation is primarily thermal in origin. The discovery of radio galaxies, quasistellar objects, and exploding galactic nuclei has shown that nonthermal energy sources are responsible for much of the total radiation emitted in the universe as well as sources of high-energy particles, x rays, and probably gravitational radiation. Chapter 12 includes a discussion of normal galaxies (that is, those whose emitted radiation is primarily thermal), which is followed by a discussion of Seyfert galaxies, radio galaxies, and quasistellar objects whose emitted radiation is primarily nonthermal.

The subject of cosmology (see Chapter 13) has always captured the imagination of philosophers and scientists. Unfortunately, hard evidence in support of cosmological theories has been scant. Chapter 13 includes a discussion of recent efforts to obtain evidence about cosmology (magnitude-redshift measurements of galaxies, numbers counts of radio sources, and background radiation measurements), in addition to an introduction to cosmological theory. Since the discovery of the

v

microwave background radiation during the last decade is the most important known clue concerning the origin and evolution of the universe, much of Chapter 13 is devoted to a discussion of its consequences.

The references at the end of this book are by no means complete. They are not included to give credit for results but rather to help the reader find more complete discussions of various topics.

A number of colleagues have been very helpful during the writing of this book. I would particularly like to thank: J. Altman, H. Bradt, S. Feldman, E. Gavrin, E. Levy, R. Rood, R. Smith, B. Tinsley, and W. Watson.

William K. Rose

College Park, Maryland
January 1973

CONTENTS

ASTROPHYSICS

CHAPTER 1
Matter
and Energy
in Space

1–1 RECESSION OF GALAXIES AND HUBBLE CONSTANT

Most observed matter is in the form of galaxies, massive aggregates of stars. Although distributed throughout space in a highly irregular manner, galaxies are generally members of groups or clusters. However, if averaged over sufficiently large distances, galaxies appear to be distributed in a more or less homogeneous and isotropic manner. For this reason, the assumption of homogeneity and isotropy is generally made in constructing cosmological models. In addition to self-gravitating bodies such as galaxies and stars, space is pervaded by other forms of energy—gas, dust, photons, high-energy particles, neutrinos, and magnetic fields. This is undoubtedly the case because some measured forms of energy (for example, photons) have very large mean free paths in space. The observed isotropy of the microwave background radiation[1] and the diffuse x-ray background is the best evidence for the large-scale isotropy of the universe.

The sky is dark except in the vicinity of stars. This elementary observation is of fundamental importance to cosmology. If space were Euclidean and static; and the large-scale distribution of stars uniform, the average energy density of radiation throughout space would eventually become equal to its value on the surface of stars. This circumstance follows from the requirements of thermodynamic equilibrium and leads to an apparent dilemma known as Olbers' paradox, which was pointed out in the early part of the nineteenth century. The resolution of Olbers' paradox was provided by the observed apparent recession of galaxies. The observations show that the mean brightnesses of cluster galaxies is related to their mean radial velocities as inferred

---

[1] Also called 3°K background radiation.

1

by measured redshifts. Observations of relatively nearby galaxies demonstrate that there is a linear velocity-distance relation, that is

$$V = H \cdot D \tag{1–1}$$

where H, the Hubble constant, is given in km-sec^{-1}-Mpc^{-1},[2] the velocity V in km/sec, and the distance D in Mpc. The observed velocity-distance relation is shown in Figure 1–1.

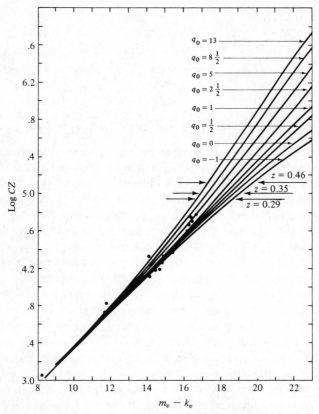

Figure 1–1   The theoretical redshift-magnitude relation. Magnitudes are on the standard visual system. Data for 18 clusters of galaxies are plotted as given by Humason, Mayall, and Sandage. Arrows are placed at the observed redshift values for three distant clusters whose magnitudes are not yet available.

Source: From Sandage, A. 1961, *Ap. J.*, **133**, 355, by permission of the University of Chicago Press.

---

[2] A parsec (pc), which is defined to be the distance from which an object of diameter equal to one astronomical unit (AU) would subtend an angle of one second of arc, is equal to 3.1×10^{18} cm. The astronomical unit is the semimajor axis of the orbit of the earth. 1 Mpc $\equiv 10^6$ pc.

In order to determine the Hubble constant, it is necessary to establish an extragalactic distance scale. The determination of the distances to external galaxies with large redshift involves several stages. Since uncertainties are associated with each stage, it is essential that independent methods of determination be employed so that the internal self-consistency of the results can be used to estimate their uncertainty. The astronomical unit (AU) is the fundamental astronomical unit of length. Measurements of the angular motions of stars in conjunction with the measured value of the AU constitute the first stage in determining the extragalactic distance scale. The distance to a cluster of stars, called the Hyades cluster, provides a calibration of the absolute magnitude of main sequence stars as a function of their color. This calibration makes it possible to estimate the distance to clusters that contain pulsating stars. The empirically determined period-luminosity relations for these stars are then used to determine the distances to neighboring galaxies. Observations of bright stars in these galaxies indicate that the brightest stars are of comparable luminosity and therefore can be used to calibrate the distances to more distant galaxies in relatively nearby clusters of galaxies, most notably Virgo. The inferred distance to the Virgo cluster determines a value for the Hubble constant. Recent determinations of H range from 50 to 125 km/sec/Mpc with 75 km/sec/Mpc as the preferred value. The inverse of the Hubble constant ($\sim 13 \times 10^9$ years) provides an estimate of the age of the universe. It is of considerable interest to compare $1/H$ with the inferred ages of old stars (see Chapter 2). The brightest galaxies in rich clusters (that is, giant ellipticals) appear to be of comparable luminosity. This circumstance makes it possible to estimate distances to very distant clusters if it is assumed that distant galaxies are of a similar nature to those observed locally. It would be of considerable interest to measure a deviation from the linear velocity-magnitude relation for galaxies with large redshifts. However, observations have so far not been of sufficient accuracy to reliably make this determination. A more complete discussion of the extragalactic distance scale will be given in Chapter 13.

1–2 NEWTONIAN COSMOLOGY AND THE DECELERATION PARAMETER

It is widely believed that the overall development of the universe is described by one of a set of cosmological models known as Friedmann models. If this is the case, then the primary goal of cosmology is to determine which model is correct. The Friedmann models are based on the assumption of isotropy and homogeneity of mass energy in space in addition to the field equations of general relativity. However, in the case of a universe where the effects of pressure can be neglected, we can derive the Friedmann models from Newtonian theory.

Let the point O in Figure 1–2 be chosen as the origin. As seen from the origin, the Hubble law for the recession of galaxies implies that all distant points (say P) appear to be expanding from O with velocity

$$\frac{d\mathbf{r}}{dt} = \mathbf{v} = H\mathbf{r} \tag{1-2}$$

where $\mathbf{r}$ is the distance from the origin to P. Transforming the origin to an arbitrary point O', we obtain

$$\mathbf{r}' = \mathbf{r} - \mathbf{r}_O$$
$$\mathbf{v}' = \mathbf{v} - \mathbf{v}_O = H(\mathbf{r} - \mathbf{r}_O) = H\mathbf{r}'. \tag{1-3}$$

Equation (1–3) shows that the universe appears the same for all observers except for local irregularities.

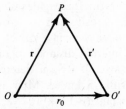

Figure 1–2   The distances between an arbitrary point P and two observers O and O' are shown. The figure shows that if a Hubble law applies for an observer O, it also applies for an arbitrary observer O'.

If a sphere of uniform density and mass

$$M = \tfrac{4}{3}\pi\rho(t)R^3(t) \tag{1-4}$$

is expanding, the equation of motion is[3]

$$\frac{d^2R}{dt^2} = -\frac{GM}{R^2}. \tag{1-5}$$

Multiplying Equation (1–5) by dR/dt and integrating, we find

$$\frac{1}{2}\left(\frac{dR}{dt}\right)^2 - \frac{GM}{R} = E \quad (E = \text{constant}). \tag{1-6}$$

If $E < 0$, then the sign of the right-hand side of the equation will change after a sufficient interval of time, and the universe will not expand indefinitely, but fall back on itself after a finite time. In this case the universe is said to be closed. If $E > 0$, the universe will expand indefinitely and, therefore, is said

---

[3] Because of spherical symmetry, mass outside radius R has no gravitational effect on mass interior to radius R.

to be open. The special case $E = 0$ corresponds to the Einstein-de Sitter model shown in Figure 1–3. Equations (1–2), (1–4), and (1–6) imply that the universe will be closed if its mean density exceeds the critical density

$$\rho_{\text{critical}} = \frac{3}{8\pi} \frac{H^2}{G} \qquad (1\text{–}7)$$

where H is in units of $\sec^{-1}$. For the present epoch, we find $\rho_{\text{critical}} \simeq 10^{-29}$ g-cm^{-3} if $H = 75$ km sec^{-1} Mpc^{-1}. In cosmology it is convenient to define a parameter q called the deceleration parameter, which is given by the expression

$$q \equiv \frac{-(d^2R/dt^2)R}{(dR/dt)^2} = \frac{-(d^2R/dt^2)}{H^2R}. \qquad (1\text{–}8)$$

It follows from Equations (1–5), (1–6), and (1–8) that the above condition for the universe to be closed (that is, $E < 0$) is equivalent to the requirement that q exceed $\frac{1}{2}$.

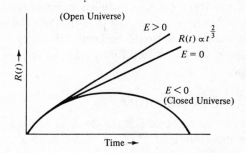

Figure 1–3   The expansion parameter $R(t)$ is shown as a function of time for an open universe ($E > 0$), a closed universe ($E < 0$) and an Einstein-de Sitter universe ($E = 0$).

In order to determine how R varies as a function of t, it is convenient to introduce a parameter τ defined by the expression

$$d\tau = \sqrt{2|E|} \, \frac{dt}{R(t)}. \qquad (1\text{–}9)$$

If $E < 0$, the parametric equations that relate R and t (the world time) are

$$R = \left(\frac{GM}{2|E|}\right)(1 - \cos \tau)$$

$$\pm (t - t_0)\sqrt{2|E|} = \frac{GM}{2|E|}(\tau - \sin \tau). \qquad (1\text{–}10)$$

The above equations describe a cycloid. For $E > 0$, we obtain

$$R = \left(\frac{GM}{2E}\right)(\cosh \tau - 1)$$

$$\pm(t - t_0)\sqrt{2E} = \left(\frac{GM}{2E}\right)(\sinh \tau - \tau)$$

(1–11)

where t_0 is some arbitrary initial time. The validity of Equations (1–10) and (1–11) can be checked by integrating Equation (1–6). Figure 1–3 describes the above solutions. The importance of radiation (and therefore the effect of radiation pressure) during the early stages of the evolution of the universe is indicated by the 3°K background radiation. Friedmann models that include the effect of pressure will be discussed in Chapter 13.

Although the results are still very inconclusive, the most recent observational evidence obtained by means of galaxy redshift measurements favors a Friedmann model with $q > \frac{1}{2}$ (that is, a closed universe). On the other hand, the most widely accepted estimate of the observed matter in galaxies is $\sim 5 \times 10^{-31}$ g/cm^3. If this estimate included the entire mass of the universe, the mean density of matter would be less than the critical density by a factor of about 20, and the universe would be open. However, it is possible that matter might exist in yet undetected forms such as very condensed objects or intergalactic gas.

1–3 INTERGALACTIC MEDIUM

For the intergalactic gas to be sufficiently dense to imply a closed universe, its mean density must be $\gtrsim 10^{-29}$ g/cm^3 (that is, $\gtrsim 5 \times 10^{-6}$ hydrogen atoms/cm^3). The feasibility of detecting such an intergalactic gas depends on its temperature. If no additional heating mechanisms were operative, the expansion of the universe would rapidly cool the intergalactic gas, and consequently intergalactic hydrogen would be neutral. However, unsuccessful attempts to detect neutral hydrogen have placed rather low limits on the content of neutral hydrogen in intergalactic space. These measurements indicate that, if the density of hydrogen is greater than 10^{-6} atoms-cm^{-3}, it must be almost completely ionized, which implies that its temperature exceeds 10^4°K. If intergalactic hydrogen is ionized and its density is greater than ρ_{crit}, then measurements of the diffuse x-ray background place an upper limit on its temperature. The precise temperature depends on the Hubble constant and the chosen cosmological model. The x-ray measurements together with the upper limit on the amount of neutral hydrogen indicate that, if the density of the intergalactic medium is to exceed the critical density, its temperature must lie in the range $10^4 \leq T \leq 10^7 - 10^8$°K.

A cold gas gets heated very rapidly until its temperature reaches 10^4°K. At this temperature it encounters a thermal barrier that arises because of strongly temperature-dependent cooling by inelastic electron collisions with atomic hydrogen. The primordial abundance of helium is expected to be ~ 30 percent by mass if the 3°K background radiation is a remnant of the big bang (see Chapter 13). The presence of appreciable helium in intergalactic space would create another thermal barrier at $\sim 5 \times 10^{4}$°K due to electron collisions with He^{+}. For temperatures in excess of 10^{5}°K, the heating of the intergalactic gas should be relatively rapid.

Other forms of energy in addition to galaxies, stray star clusters, and gas are undoubtedly distributed throughout space. A summary of the estimated energy content of these constituents of space is given in Table 1–1. The average observed energy density of observed matter in galaxies (~ 300 eV/cm^3) is seen to be much greater than the estimated energy densities of the probable constituents of space that are listed in Table 1–1.

Table 1–1

Constituent	Energy Content (average all space including intergalactic space)	Primary Sources
A. Photons		
1. Infraradio	?	?
2. Radio and microwave (exclude 3°K radiation)	1μ eV/cm^3	Galaxy, radio galaxies, quasars
3. 3°K background	0.4 eV/cm^3	Big bang
4. Infrared	? may be $>$ 3°K radiation	Seyfert galaxies, quasars, N galaxies
5. Visible light	0.01 eV/cm^3 (0.3 eV/cm^3 in galaxy)	Stars in normal galaxies
6. Soft x rays (<1 keV)	50–100μ eV/cm^3	Intergalactic plasma?
7. Hard x rays (1–500 keV)	100μ eV/cm^3	Most likely inverse Compton scattering of fast electrons with background radiation; discrete sources
8. Gamma rays	?	?
B. Magnetic fields	0.001 eV/cm^3?	?
C. Cosmic rays and fast electrons	100μ eV/cm^3	Radio galaxies, quasars, supernovae in normal galaxies
D. Neutrinos	0.1 eV/cm^3	Big bang
E. Gravitons	0.1 eV/cm^3	Big bang

Bremsstrahlung

Hot, ionized gases are common in nature and consequently bremsstrahlung, which is produced by the interaction of free electrons with ions, is an extremely important radiative process.

The well-known classical, nonrelativistic expression for the power radiated by an accelerated charge is

$$P(t) = \frac{2}{3} \frac{e^2}{c^3} a(t)^2 \qquad (1\text{-}12)$$

where $a(t)$ is the acceleration of the charge as a function of time. The total energy radiated by an electron moving past an ion is

$$E = \int_{-\infty}^{+\infty} P(t)\, dt = \frac{2}{3} \frac{e^2}{c^3} \int_{-\infty}^{+\infty} a^2(t)\, dt. \qquad (1\text{-}13)$$

The radiation from an electron moving in the Coulomb field of an ion (see Figure 1-4) can be estimated by means of the following physical model.

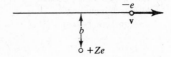

Figure 1-4   An electron of change $-e$ passes an ion of charge Ze. b is the impact parameter (distance of closest approach).

The acceleration experienced by an electron at a distance $\mathbf{r}$ from an ion of charge Ze is

$$\mathbf{a} = -\frac{Ze^2 \mathbf{r}}{r^3 m} \qquad (1\text{-}14)$$

where m is the mass of the electron. If b is the impact parameter (that is, the distance of closest approach) and v is the velocity of the electron, then the Coulomb force acts strongly for a time of order b/v, and consequently the frequency of a photon emitted by an electron with impact parameter b is

$$v \sim \frac{v}{2\pi b}. \qquad (1\text{-}15)$$

It follows from Equations (1-13) and (1-14) that the energy radiated by each electron as it passes an ion is approximately

$$E \sim \frac{2}{3} \frac{Z^2 e^6}{m^2 c^3 b^3 v} \text{ ergs} \qquad (1\text{-}16)$$

where it has been assumed that the electron is accelerated by a constant amount for a time b/v. If the electron flux is assumed equal to unity (that is,

$n_e v = 1$), the energy radiated per unit time per ion by means of electrons with impact parameters between b and $b + db$ and velocity v is

$$E_v \, dv = E 2\pi b \, db \sim \frac{8\pi^2 Z^2 e^6 \, dv}{3m^2 c^3 v^2}. \tag{1-17}$$

Equation (1–17) cannot be correct at high frequencies because electrons cannot emit photons with energies that exceed the initial kinetic energy of the electron, and therefore

$$v_{max} = \frac{\frac{1}{2}mv^2}{h} \tag{1-18}$$

is the high-frequency cutoff above which the radiated power drops to zero. The approximate result given in Equation (1–17) is equal to the exact classical result except for a factor of order unity. It would appear likely that large deviations from the classical results should arise at temperatures such that kT is greater than the ionization potential of hydrogen. However, calculations based on quantum theory show that the classical result given above is nearly correct so long as kT is less than the rest mass energy of the electron (that is, $T \lesssim 6 \times 10^{9}°K$).

If the velocity distribution of the electrons is a Maxwell-Boltzmann distribution, the probability that an electron be in an element of velocity space d^3v is

$$f(v) \, d^3v = \left(\frac{m}{2\pi kT}\right)^{3/2} \exp\left(\frac{-mv^2}{2kT}\right) d^3v \tag{1-19}$$

where

$$\int f(v) \, d^3v = 1.$$

Summing over the velocity distribution defined in Equation (1–19), it follows from Equation (1–17) that the total power per cm^3 radiated in the frequency range v to $v + dv$ is

$$4\pi j_v \, dv = n_e n_i \left[\int_{v_{min}}^{\infty} f(v) v E_v \, d^3v \right] dv \tag{1-20}$$

where

$$v_{min} = \left(\frac{2hv}{m}\right)^{1/2}$$

is the minimum velocity for an electron to radiate a photon of frequency v. Integrating Equation (1–20) over velocity space, we find

$$4\pi j_v \, dv = \frac{8\pi}{3mc^3} \left(\frac{2\pi}{kTm}\right)^{1/2} Z^2 e^6 n_e n_i e^{-(hv/kT)} \, dv. \tag{1-21}$$

The total power radiated is found by integrating Equation (1–21) over frequency and multiplying by a numerical factor to make the result consistent with more exact calculations. We find

$$4\pi \int j_v \, dv = 1.4 \times 10^{-27} Z^2 T^{1/2} n_e n_i \quad (\text{erg cm}^{-3} \text{ sec}^{-1}). \quad (1\text{--}22)$$

This classical result agrees with quantum mechanical calculations to within a numerical factor called the gaunt factor, which is of order unity.

1–4 CONTENTS OF GALAXY

A schematic model of the galaxy is shown in Figure 1–5. Because the galactic disk is rapidly rotating, the galactic system balances gravity by means of centrifugal force rather than pressure as is the case for stars. The major known constituents of interstellar space are shown in Table 1–2. Figure 1–6 shows the density distribution of neutral and ionized hydrogen in the galactic plane as a function of distance from the galactic center. There is an observed deficiency in the quantity of neutral and ionized gas within a distance of about 3 kpc from the galactic center. Similar results are indicated for other galaxies. The peak in the galactic neutral hydrogen density distribution shown in Figure 1–6 occurs beyond that for ionized hydrogen, whose presence is indicative of star formation. For this reason, the occurrence of star formation is apparently not a simple function of neutral gas density.

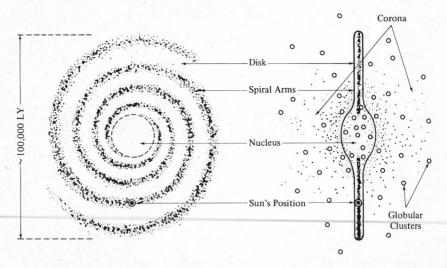

Figure 1–5   A schematic representation of some of the major components of the galaxy.

Table 1–2

Constituents of Interstellar Space	Average Mass or Energy Content
Stars (observed)	$0.075 \ M\odot/(pc)^3$
Population I	$\sim 0.06 \ M\odot/(pc)^3$
Population II	$\sim 0.015 \ M\odot/(pc)^3$
Gas—mostly neutral hydrogen and helium	$\sim 0.025 \ M\odot/(pc)^3$
Dust—radii of grains $\sim 10^{-5}$ cm	$\sim 0.0002 \ M\odot/(pc)^3$
Cosmic rays	$\sim 0.5 \ eV/cm^3$
Magnetic fields—$H \sim 3 \times 10^{-6}$ G	$\sim 0.2 \ eV/cm^3$
Starlight	$\sim 0.5 \ eV/cm^3$

Most of the energy density within our galaxy is in the form of stars. Many stars are known to be grouped into clusters, which range from rich aggregates of stars that are gravitationally bound to loose associations that are gravitationally unbound and therefore dispersing into the interstellar medium. Because individual stars in clusters are at approximately the same distance, stellar clusters are very important in helping to compare the relative luminosities of stars with differing surface temperatures. The ages and initial chemical compositions of stars in a single cluster are probably similar since

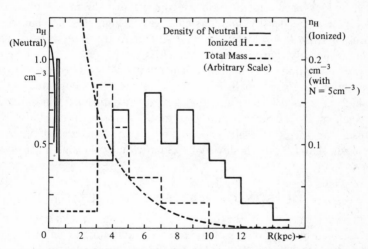

Figure 1–6   Smoothed density distribution as a function of distance to the galactic center, R, in the galactic plane, for neutral hydrogen, ionized hydrogen (assuming all ionized clouds have $N = 5 \ cm^{-3}$), and total mass.

Source: From Kerr, F. and G. Westerhout. 1965. *Galactic Structure*, eds. Blaauw and Schmidt. Chicago: University of Chicago Press. Reprinted by permission of the University of Chicago Press.

they were probably formed from the same interstellar material at about the same time. Globular clusters are spherical stellar systems of about 10^5 M$\odot$ that are found far from the galactic plane and do not follow the rotation of the galaxy.[4] About 10^2 of these stellar systems are known to exist in our galaxy. Galactic clusters (often called open clusters) contain far fewer stars than globular clusters and are much less centrally condensed. They are found in the disk of the galaxy and probably originate in spiral arms. About 10^3 galactic clusters, many of which contain young, massive stars, are observed in our galaxy. However, because galactic clusters are concentrated near the galactic plane, where they can be readily obscured by interstellar dust, it is likely that we observe only a small fraction of these stellar systems. The most luminous stars in our galaxy are known to be grouped into gravitationally unbound stellar associations ($M \sim 10^2$ M$\odot$) that are found along the spiral arms.

It is convenient to divide stars into two major groups based on their observed kinematic properties. Population II stars that are in the solar neighborhood have relatively high space velocities ($\gtrsim 60$ km/sec) relative to the local standard of rest, which is defined by the requirement that the local mean velocity of stars equal zero. These high-velocity stars are very old and include many red giants. They are similar to the stars found in globular clusters, which also have high space velocities. Local Population I stars have relatively low space velocities with respect to the sun. This population contains nearly all the young, massive stars. It is generally found that young stars have low space velocities and are concentrated toward the plane of the galaxy where the density of interstellar gas is highest. The characteristic features of these low-velocity stars are similar to those of stars in galactic clusters, which also have low space velocities.

After formation, stars maintain their initial velocities because collisional cross sections are very small. It is widely believed that Population II stars have high space velocities because they were formed during the initial collapse of the galaxy. On the other hand, Population I stars were formed after initial collapse, and therefore their velocities reflect the velocities of interstellar gas clouds that have been slowed by cloud-cloud collisions and turbulent dissipation.

1–5 LOCAL DENSITY OF MATTER (OORT LIMIT)

As will be described below, the total density of mass in the solar neighborhood as determined by the motions of stars in the gravitational field of the galaxy is about 0.15 M$\odot$/(pc)^3. This value for the total mass density is about 50 percent greater than the density in the form of observed stars and gas in the solar neighborhood (see Table 1–2). We assume that the galaxy is

---

[4] M$\odot$ denotes the mass of the sun (M$\odot$ = 1.99×10^{33}g).

stratified in plane parallel layers with z perpendicular to the plane of the galaxy. The energy of a star associated with motion in the z direction is a constant of the motion and can be written

$$E = \frac{p_z^2}{2M} + M\Phi(z) \tag{1-23}$$

where M is the mass of the star, p_z its momentum in the z direction, and $\Phi(z)$, the gravitational potential, is a function of the distance from the plane of the galaxy only. We define the distribution function f so that

$$f(E)\, dp_z\, dV \tag{1-24}$$

gives the number of stars within dp_z and the volume element dV. The mean square momentum in the z direction is

$$\langle p_z^2 \rangle = \frac{1}{n} \int f(E) p_z^2\, dp_z \tag{1-25}$$

where n, the number of stars per unit volume, is

$$n = \int_{-\infty}^{+\infty} f(E)\, dp_z. \tag{1-26}$$

If we multiply Equation (1–25) by n and differentiate with respect to z, we find

$$\frac{\partial}{\partial z}(n\langle p_z^2 \rangle) = -Mg_z(z) \int \frac{df(E)}{dE} p_z^2\, dp_z \tag{1-27}$$

where we have used the relation

$$\frac{\partial E}{\partial z} = M\frac{d\Phi(z)}{dz} = -Mg_z(z). \tag{1-28}$$

Integrating by parts, it follows that

$$\frac{1}{n}\frac{d}{dz}\left(\frac{n\langle p_z^2 \rangle}{M^2}\right) = -g_z(z). \tag{1-29}$$

The above equation shows that, since the velocities ($\langle v_z^2 \rangle = \langle p_z^2 \rangle / M^2$) and spatial distribution of stars of similar mass can be deduced from observations, $g_z(z)$ can be determined. Once $g_z(z)$ has been obtained, $dg_z(z)/dz$ is determined, and consequently Poisson's equation,

$$\frac{dg_z(z)}{dz} = -4\pi G\rho(z), \tag{1-30}$$

can be used to find the spatial density of matter in the solar neighborhood.

At one time it was believed that the difference between the total mass and the mass of observed stars and gas might be in the form of H_2 molecules.

Although recent attempts to detect H_2 have been successful, its observed abundance is inadequate to explain the missing mass. It is possible that the missing mass is in the form of yet undetected forms of matter such as low-mass stars, collapsed stars, or baseball-size particles.

1–6 NEUTRAL HYDROGEN

A hydrogen atom consists of an electron moving in the Coulomb field of a proton. Its physical properties are determined by the Schrödinger wave equation:

$$-\frac{\hbar^2}{2m_r}\frac{1}{r^2}\frac{\partial}{\partial r}\left(r^2\frac{\partial\psi}{\partial r}\right) + V(r)\psi = E\psi \tag{1-31}$$

with

$$V(r) = -\frac{e^2}{r} + \frac{l(l+1)\hbar^2}{2m_r r^2}$$

and

$$m_r = \frac{mm_p}{m+m_p}$$

where l is the orbital angular momentum, and m_r is the reduced mass of the electron. The lowest energy state of the hydrogen atom has $l = 0$. Rewriting Equation (1–31) and placing $l = 0$, it follows that

$$\frac{\hbar^2}{2m_r}\left(\frac{\partial^2\psi}{\partial r^2} + \frac{2}{r}\frac{\partial\psi}{\partial r}\right) + \left(E + \frac{e^2}{r}\right)\psi = 0. \tag{1-32}$$

Substituting

$$\psi(r) = e^{-(r/a_0)}$$

into Equation (1–32), it follows that the equation will be satisfied for all values of r if

$$a_0 = \frac{\hbar^2}{m_r e^2}$$

and

$$E = -\frac{\hbar^2}{2m_r a_0^2} = -\frac{m_r e^4}{2\hbar^2} = 13.6 \text{ eV.}$$

The quantity a_0, which is called the Bohr radius, is the most probable distance of the electron from the proton, and E is the energy of the ground state of the hydrogen atom. The remaining energy levels are given approximately by the familiar expression

$$E_n = -\frac{m_r e^4}{2\hbar^2 n^2} = \frac{13.6}{n^2} \text{ eV} \qquad (n = 1, 2, 3, 4, \ldots). \tag{1-33}$$

In regions of interstellar space that are predominantly neutral hydrogen, the kinetic energy of the atoms is insufficient to collisionally excite levels with

$n > 1$, and consequently emission lines from such excited levels cannot be observed. However, absorption caused by the Lyman α transition can be seen against the ultraviolet (UV) spectra of hot stars.

The hydrogen ground state is itself degenerate because, if a coordinate axis is defined, an electron can orient its spin in either the positive or negative direction, and likewise the spin of the proton can point along either direction. The degeneracy of the hydrogen ground state is split by the interaction of the magnetic moments of the electron and proton. Because of this interaction, the $S = \frac{1}{2}$ electron couples with the $I = \frac{1}{2}$ proton to form a hyperfine doublet: $F = 0$ and 1.[5] The energy splitting of this doublet is

$$hv = \tfrac{8}{3} g_p \alpha^2 R_y \tag{1–34}$$

where $\alpha = e^2/\hbar c$ is the fine structure constant; g_p, the g factor of the proton, equals 5.58 (m/m_p); and R_y, the Rydberg, is equal to 13.6 eV. The frequency given in Equation (1–34) is 1.42×10^9 Hz and corresponds to a wavelength of 21 cm. Since the orbital angular momentum of both the $F = 1$ and $F = 0$ states of the doublet is zero, transitions between these levels violate the $\Delta l = \pm 1$ selection rule. This implies that electric dipole radiation is forbidden, and consequently magnetic dipole radiation is the lowest-order radiative process. The transition probability for spontaneous emission between the $F = 1$ and $F = 0$ levels is

$$A_{10} \simeq 3 \times 10^{-15} \text{ sec}^{-1} \simeq 10^{-7} \text{ yr}^{-1}. \tag{1–35}$$

The long lifetime of the 21-cm transition arises not only because it is of the magnetic dipole type but also because of the low frequency of the transition (since A_{10}, the Einstein coefficient for spontaneous emission discussed in Appendix A, varies as v^3).

In order to discuss the absorption or emission of 21-cm line radiation, it is necessary to derive an equation that describes the transfer of this radiation. Let dA be an infinitesimal area shown in Figure 1–7 at position $\mathbf{r}$. The unit vector $\hat{n}$ is normal to dA. Let dE_v be the radiant energy flowing through dA in the frequency interval between v and $v + dv$ and in the solid angle $d\omega$ about direction θ. The intensity of radiation is defined as

$$I_v(\mathbf{r}, \theta, t) = \lim_{\substack{dA, dt \\ d\omega, \\ dv \to 0}} \frac{dE_v}{dA \cos \theta \, d\omega \, dv \, dt} \qquad \text{ergs cm}^{-2} \text{ steradian}^{-1} \text{ Hz}^{-1} \text{ sec}^{-1}.$$

$$\tag{1–36}$$

The mean intensity J_v is

$$J_v(r, t) = \frac{1}{4\pi} \int I_v \, d\omega = \frac{c}{4\pi} U_v \tag{1–37}$$

where U_v is the radiation energy density per unit frequency interval.

---

[5] $F = S + I, S + I - 1, \cdots |S - I|$. The degeneracy of the $F = 1$ and $F = 0$ states are $2F + 1 = 3$ and 1 respectively.

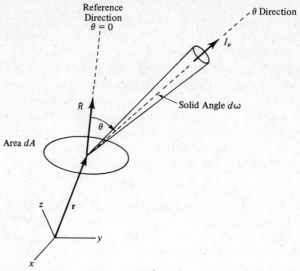

Figure 1-7 The intensity of radiation I_v at an angle θ with respect to the normal to the area element dA.

If we neglect scattering, the change in the intensity of radiation along an element of path length ds is

$$dI_v = [-k_v I_v + j_v]\, ds \qquad (1\text{-}38)$$

where k_v is the absorption coefficient, and j_v is the emission in units of ergs-sec^{-1}-cm^{-3} steradian^{-1}-Hz^{-1}. The first term inside the brackets represents the amount of radiation per unit length that is absorbed by the medium through which the radiation is transversing. The second term represents the corresponding amount of radiation emitted by the medium. The optical depth $d\tau_v$ along a path length ds is defined by the expression

$$d\tau_v = k_v\, ds. \qquad (1\text{-}39)$$

Equations (1-38) and (1-39) can be combined to produce the equation of radiative transfer:

$$\frac{dI_v}{d\tau_v} = -I_v + \frac{j_v}{k_v}. \qquad (1\text{-}40)$$

The solution to Equation (1-40) is

$$I_v = I_v(0)e^{-\tau_v} + e^{-\tau_v}\int_0^{\tau_v} \frac{j_v}{k_v}\, e^{\tau'_v}\, d\tau'_v \qquad (1\text{-}41)$$

where $I_v(0)$ is the intensity of radiation at a source located at position 0, and the integration is carried out along the path between the source and the observer.

For the 21-cm transition, j_v becomes

$$j_v = \frac{hv}{4\pi} n_1 A_{10} \phi_v \tag{1-42}$$

where n_1 is the number density of hydrogen atoms in the $F = 1$ state; A_{10} is the transition probability $(\sec^{-1})$ for spontaneous emission between states $F = 1$ and $F = 0$; and ϕ_v is the line shape factor, which is normalized by the condition

$$\int_0^\infty \phi_v \, dv = 1. \tag{1-43}$$

If the velocity distribution of the hydrogen atoms is a Maxwell-Boltzmann distribution at temperature T, the probability that the speed of a hydrogen atom lies between v and $v + dv$ is

$$4\pi f(v) v^2 \, dv = 4\pi \left(\frac{m_H}{2\pi kT} \right)^{3/2} e^{-(1/2)(m_H v^2/kT)} v^2 \, dv \tag{1-44}$$

where

$$4\pi \int_0^\infty f(v) v^2 \, dv = 1,$$

m_H is the mass of the hydrogen atom, and T is the temperature of the gas. The Doppler relation between the frequency shift $v - v_0$ and the velocity v is

$$\frac{v - v_0}{v_0} = \frac{v}{c} \cos \theta \tag{1-45}$$

where $v \cos \theta$ is the projection of the velocity of the atom along the direction between the atom and the observer. Equations (1–44) and (1–45) imply that for the case of a Doppler broadened line, the line shape factor ϕ_v is

$$\phi_v = \sqrt{\frac{m_H c^2}{2\pi kT v_0^2}} \exp \left[-\frac{m_H c^2 (v - v_0)^2}{2kT v_0^2} \right]. \tag{1-46}$$

The line width of 21-cm line radiation is due primarily to the turbulent motions of the gas. These motions produce a spread in the observed line width that far exceeds the Doppler width caused by the thermal motions of the gas.

In order to obtain a solution to the equation of radiative transfer, it is necessary to derive an expression for the absorption coefficient k_v. For the time being we assume that the medium is in thermodynamic equilibrium. This assumption allows us to use the principle of detailed balance, which asserts that each physical process must be exactly balanced by its inverse under conditions of thermodynamic equilibrium. Consequently the amount of energy absorbed must equal the amount emitted at each frequency, that is,

$$j_v = k_v B_v \qquad (1\text{-}47)$$

where the blackbody function is

$$B_v(T) = \frac{2hv^3}{c^2} \frac{1}{[\exp(hv/kT) - 1]}. \qquad (1\text{-}48)$$

Equation (1-47) is a statement of Kirchhoff's law. It follows from Equations (1-42) and (1-47), and the definitions of the Einstein coefficients given in Appendix A that the absorption coefficient k_v is

$$k_v = \frac{hv}{c} \phi_v B_{10} n_1 \left[\exp\left(\frac{hv}{kT}\right) - 1 \right] \qquad (1\text{-}49)$$

where n_0 and n_1 are the number densities of hydrogen atoms in the $F = 0$ and $F = 1$ states, respectively, and B_{01} (and B_{10}) are the Einstein coefficients for absorption (and induced emission). Since $B_{10} = (g_0/g_1)B_{01}$ with $g_0 = 1$ and $g_1 = 3$, Equation (1-49) reduces to

$$k_v = \frac{hv}{c} n_0 B_{01} \phi_v \left[1 - \frac{n_1}{3n_0} \right]. \qquad (1\text{-}50)$$

If the optical depth τ_v is very large and the temperature uniform, Equation (1-41) becomes

$$I_v = \frac{j_v}{k_v}. \qquad (1\text{-}51)$$

From Equations (1-50), (1-51), (1-42) and the relation $A_{10} = 8\pi hv^3 B_{10}/c^3$ derived in Appendix A, we find

$$\frac{j_v}{k_v} = \frac{2hv^3/c^2}{(3n_0/n_1) - 1}. \qquad (1\text{-}52)$$

If a spin temperature T_s is defined by means of the Boltzmann relation,

$$\frac{n_1}{n_0} = \frac{g_1}{g_0} \exp\left(-\frac{hv}{kT_s}\right) \qquad (1\text{-}53)$$

the right-hand side of Equation (1-52) becomes equal to the blackbody function $B_v(T_s)$. Measured values of $I_v = j_v/k_v$ imply that $T_s \sim 100°K$ in regions of interstellar space for which $\tau_v \gg 1$. Since $hv \ll kT_s$, we find

$$n_1 = 3n_0 = \tfrac{3}{4}n_H \qquad (1\text{-}54)$$

where n_H is the total number density of hydrogen atoms.

On the other hand, if $\tau_v \ll 1$ and $I_v(0) = 0$, then from Equations (1–41) and (1–42) we find

$$I = \int I_v \, dv = \frac{h\nu}{4\pi} A_{10} \int n_1 \, ds. \tag{1–55}$$

Equation (1–55) implies that measured values of I can be used to determine the number of hydrogen atoms per square centimeter along the line of sight. These measurements indicate that

$$\bar{n}_H \sim 0.5 \text{ cm}^{-3}$$

in the galactic plane.

If we assume Kirchhoff's law ($j_v = k_v B_v$) to be valid and also assume that the spin temperature T_s is uniform throughout the region between the source and an observer, Equation (1–41) reduces to

$$I_v = I_v(0)e^{-\tau_v} + B_v(T_s)(1 - e^{-\tau_v}). \tag{1–56}$$

At radio wavelengths, the Rayleigh-Jeans approximation is valid, and therefore

$$I_v = \frac{2v^2 kT}{c^2}. \tag{1–57}$$

Equations (1–56) and (1–57) imply

$$T_b = Te^{-\tau_v} + T_s(1 - e^{-\tau_v}) \tag{1–58}$$

where T_b is the measured brightness temperature, and T is the temperature of the source. Equation (1–58) shows that 21-cm line radiation is seen in absorption if the continuum background source is of higher temperature than the foreground neutral hydrogen gas. Otherwise, it is seen in emission.

In the above discussion we have assigned a hydrogen spin temperature [see Equation (1–53)] and have shown that if an extended, homogeneous source is observed with a radio telescope, kT_s is the measured brightness temperature. It is important to point out that the spin temperature T_s does not necessarily reflect the kinetic temperature of the interstellar gas, but rather some value that is intermediate between the kinetic temperature T_K and the radiation temperature T_R. If the collision rate (mostly H atoms with H atoms) is sufficiently high, T_s will be close to T_K. On the other hand, as collisions become less important, T_s will approach T_R.

The value of T_R is determined by the ultraviolet and radio background. The presence of dust in interstellar space will lead to the absorption of Lyman α photons and consequently the radio background is likely to be most important in determining the radiation temperature. The equation of statistical equilibrium can be written

$$n_1(C_{10} + A_{10} + B_{10}U_{10}) = n_0(C_{01} + B_{01}U_{10}) \tag{1–59}$$

where n_1 and n_0 are the number densities of atoms in the $F = 1$ and $F = 0$

states, respectively; A_{10}, B_{10}, and B_{01} are the Einstein coefficients defined in Appendix A; U_{10} is the radiation energy density per unit frequency interval at 1.42×10^9 Hz, and C_{10} is the probability per second that a hydrogen atom that is initially in the triplet state will go to the singlet state as the result of a collision. The principle of detailed balance, which states that under equilibrium conditions the number of transitions from one state to another is balanced by the inverse rate, implies the relation

$$C_{01} = 3C_{10} \exp\left(-\frac{h\nu}{kT_K}\right). \tag{1-60}$$

Although the intensity of the radio background radiation is not likely to be that of a blackbody, we can write U_{10} in the form

$$U_{10} = \frac{8\pi h\nu^3}{c^3}\left[\exp\left(\frac{h\nu}{kT_R}\right) - 1\right]^{-1} \tag{1-61}$$

where T_R is some function of ν.

Equation (1–60) and the definitions of the Einstein coefficients given in Appendix A imply that Equation (1–59) can be rewritten as

$$n_1\left(C_{10} + A_{10} + A_{10}\frac{c^3}{8\pi h\nu^3}U_{10}\right)$$

$$= 3n_0\left[C_{10}\exp\left(-\frac{h\nu}{kT_R}\right) + A_{10}\frac{c^3}{8\pi h\nu^3}U_{10}\right]. \tag{1-62}$$

From Equations (1–53) and (1–62), we find

$$\exp\left(\frac{-h\nu}{kT_s}\right) = \frac{C_{10}\exp\left(-h\nu/kT_K\right)[\exp\left(h\nu/kT_R\right) - 1] + A_{10}}{(C_{10} + A_{10})[\exp\left(h\nu/kT_R\right) - 1] + A_{10}}. \tag{1-63}$$

Since $h\nu$ is generally much less than kT_R or kT_K, we can write

$$T_s = \frac{x}{1 + x}T_K + \frac{1}{1 + x}T_R \tag{1-64}$$

with

$$x = \frac{h\nu C_{10}}{kT_K A_{10}}.$$

1-7 THERMAL INSTABILITY AND INTERSTELLAR CLOUDS (HI REGIONS)

In the preceding section, we discussed the presence of neutral hydrogen in interstellar space and pointed out that the mean hydrogen number density is about $n_H = 0.5$ cm^{-3}. The observed spatial distribution of interstellar hydrogen is highly irregular with most of the gas concentrated in relatively dense clouds called HI regions. The densities of interstellar clouds use typically $n_H \sim 10\text{--}10^2$ cm^{-3}, and they are observed on all discernible scales ranging from $\sim 10^3$ AU to ~ 300 pc. The random motions (~ 10 km/sec) and velocity dispersions inside the clouds ($\sim 1\text{--}8$ km/sec) indicate that the interstellar medium is in a highly turbulent state. It is clear that a satisfactory theory of the interstellar gas must explain its observed clumpiness and also take into account its turbulent motions.

The energy balance that results from opposing mechanisms of energy gain and energy loss determine the equilibrium temperature of an interstellar cloud. In HI regions the dominant heating mechanism is probably the photoionization of such trace elements as C, Fe, Si, and Na by means of low-energy cosmic rays, ultraviolet radiation, or soft x rays. Ionization will heat the gas because the ejected electron will redistribute its excess kinetic energy before it is likely to recombine with an ion. Heating of the interstellar gas may also be caused by Alfven waves (see Section 3–4) that are excited by cosmic rays. The primary source of cooling is likely to be collisional excitation followed by radiative decay of excited energy levels sufficiently close to their respective ground states that they can be excited by means of particles of energy comparable to kT.

The first law of thermodynamics asserts that the amount of heat added or taken away from a mass element is

$$\frac{dU}{dt} + P\frac{d(1/\rho)}{dt} = \frac{\Gamma - \Lambda}{\rho} \tag{1-65}$$

where U is the internal energy per gram; ρ is the density; P is the pressure; and Γ (and Λ) are the heating (and cooling) rates, respectively, in units of ergs cm^{-3} sec^{-1}.

The equilibrium temperature of a volume element of interstellar gas, which is transparent to its own radiation, is determined by the condition $\Gamma - \Lambda = 0$. A plausible theory for the origin of interstellar clouds asserts that they arise as a consequence of the thermal instability of the interstellar gas. If a gas is thermally unstable, an arbitrarily small perturbation of its temperature away from its equilibrium value will lead to a spontaneous heating or cooling of the gas. If pressure equilibrium is maintained during the displacement from equilibrium, then the condition for thermal instability to arise is

$$\left[\frac{\partial(\Gamma - \Lambda)}{\partial T}\right]_P = \frac{\partial(\Gamma - \Lambda)}{\partial T} + \frac{\partial(\Gamma - \Lambda)}{\partial \rho}\left[\frac{\partial \rho}{\partial T}\right]_P > 0 \tag{1-66}$$

where it has been assumed that $\Gamma - \Lambda = f(T, \rho(T, P))$ and that the derivative with respect to temperature is taken at constant pressure.

The simplest example of a thermally unstable gas is a very hot $(T \gtrsim 10^7°K)$ plasma that is transparent to its own radiation. Bremsstrahlung is the principal emission mechanism for such a hot gas. If the heating function Γ can be neglected, then the quantity $\Gamma - \Lambda$ is proportional to $-\rho^2 T^{1/2}$ [see Equation (1–22)], and consequently from Equation (1–66) it follows that the plasma is thermally unstable. Instability arises because a temperature increase (decrease) is accompanied by a density decrease (increase), and therefore the rate of bremsstrahlung emission becomes less (more). This circumstance implies that once the temperature of the gas has been perturbed it will continue to change in the same sense.

Although the heating and cooling functions are not securely known for the interstellar medium, estimates of the functions can be made. Such estimates indicate that the gas is probably unstable and that the temperature of the unstable gas should rise to a stable value of $T \sim 10^4°K$ or fall to a stable value of $T \sim 50°K$. It has been proposed that the observed interstellar cloud structure can be understood as a quasihydrostatic equilibrium between hot and cold gas phases.

The origin of thermal instability in HI regions may be understood on the basis of the following physical model. The heating of the interstellar gas is assumed to be proportioned to n_H, the number of neutral hydrogen atoms per cubic centimeter and nearly independent of the temperature. On the other hand, the cooling rate is proportioned to $n_e n_i$, where n_e is the electron density, and n_i is the number density of ions and atoms responsible for the cooling. The temperature dependence of the cooling rate will depend on how the excited levels are distributed in energy about their ground states. Since the heating and cooling rates depend on different powers of the density, the equilibrium temperature will depend on the density as well as the assumed radiation field and cosmic ray flux. At high densities cooling will be relatively efficient, and therefore the temperature will be relatively low. It follows from Equation (1–66) that the interstellar medium will be thermally unstable if the temperature dependence of the cooling rate is sufficiently low. For example, if Γ is small and $-\Lambda$ varies as the square of the density, then for stability the temperature dependence must be greater than the first power of the temperature. For temperatures less than about 7000°K, the cooling function is likely to be caused by the 15.6-μ fine structure line of singly ionized carbon and several lines of neutral oxygen and singly ionized iron. If a single line were to dominate the cooling function, then there would be an exponential dependence on temperature because of the Maxwell-Boltzmann distribution of the particles responsible for exciting the level. Such a strong temperature dependence would imply thermal stability. However, the presence of a number of different energy levels implies a much weaker temperature dependence, and consequently thermal instability is possible.

1-8 INTERSTELLAR GRAINS

Observations of the spectra of stars located at different parts of the galaxy indicate the presence of extinction. Interstellar extinction, which is defined to include both the scattering and absorption of starlight, is confined primarily to the plane of the galaxy where most of the interstellar gas is concentrated. At visual wavelengths, the dependence of extinction on wavelength is approximately λ^{-1}. The observed polarization of starlight is also associated with interstellar extinction. It is generally accepted that interstellar extinction and polarization are caused by grains (small solid particles) that tend to be present in the form of more or less discrete clouds and are correlated with HI regions. The presence of interstellar grains is also indicated by the existence of reflection nebulae that are associated with stars whose surface temperature is less than about 15,000°K. Diffuse nebulae (see Section 1-10) are also known to contain grains. These nebulae possess strong emission lines and a relatively weak continuum. Interstellar polarization of starlight indicates that grains are preferentially aligned, probably as a result of interstellar magnetic fields (see Section 1-12).

The theoretical problem associated with the observed extinction and polarization of starlight is to explain the observed properties without requiring excessive mass densities (see Section 1-3). Scattering by means of free electrons and/or atoms and molecules can be ruled out because such scattering would require very large masses and give the wrong reddening law (no wavelength dependence for electron scattering and λ^{-4} dependence for Rayleigh scattering by means of atoms or molecules).

Although it is unlikely that interstellar grains are spherical in shape, the diffraction of light by spherical particles (Mie scattering) is of great importance since this problem can be solved exactly. The wavelength dependence of interstellar extinction tells us that the radii of the particles are likely to be comparable with the wavelength of light (that is, 10^{-5}–10^{-6} cm). If interstellar particles were much smaller than the wavelength of light, the scattering law would be the same as that of Rayleigh scattering (λ^{-4}). On the other hand, if the radii of the particles were much larger than a wavelength, the extinction law would be nearly wavelength independent unless absorption lines were associated with the grain material. The cross section for extinction of visual light by particles whose radii are comparable to a wavelength is about $0.3\pi r_g^2$, where r_g is the radius of the grain. If the density of grain material is taken to be 1 g/cm^3, then it can be shown that the mass of interstellar grains required to explain interstellar extinction is about one percent of the mass of the interstellar medium. The chemical composition of interstellar grains is unknown. However, silicates and graphite particles are two possibilities.

The temperature of interstellar grains is a consequence of the energy balance between the absorption of dilute radiation from stars and emission in the infrared. The mean interstellar radiation field corresponds to a blackbody at $T \sim 10^4°$K that is diluted by a factor $W \sim 10^{-14}$. If the emitted radiation were like a blackbody, the temperatures of the grains would be $W^{1/4}T \sim 3°$K. However, this estimate is incorrect since the radii of grains are much less than the characteristic wavelength of $3°$K radiation, and consequently they emit less effectively than blackbodies. When this factor is taken into consideration, the predicted temperature of interstellar grains is about $10°$K.

One hypothesis concerning the origin of interstellar grains asserts that they condense out of the interstellar medium. The process of grain formation involves two basic steps. The first step is nucleation, which involves the formation of a stable aggregate of atoms ($\sim 10^2$), the second the growth of these nuclei into grains. If f is the fraction of heavy atoms of mass M that stick onto the surface of a grain upon collision and n their total number density, the rate of increase of the mass of a grain of radius r and density ρ_g becomes

$$4\pi r^2 \rho_g \frac{dr}{dt} = 4\pi r^2 nf \left(\frac{kT}{2\pi M}\right)^{1/2} \tag{1-67}$$

where $(kT/2\pi M)^{1/2}$ is the mean Maxwellian velocity of the colliding atoms. If f is assumed to be 0.1, then the solution of Equation (1–67) shows that grains can develop to the required radii in less than the lifetime of the galaxy. Because grains are believed not to attain very large radii, it has been suggested that grains are continuously destroyed in the interstellar medium.

The above theory for the origin of interstellar grains can be criticized because it does not explain their nucleation. An alternate theory for their origin asserts that they are formed in the outer atmospheres of stars and then ejected into the interstellar medium.

We consider a plane electromagnetic wave of the form

$$E = E_0 \sin \omega t. \tag{1-68}$$

The polarization, which is the dipole moment per unit volume, becomes

$$P = X'(\omega)E_0 \sin \omega t + X''(\omega)E_0 \cos \omega t \tag{1-69}$$

where $X'(\omega)$ and $X''(\omega)$ represent the in-phase and out-of-phase components of the electric susceptibility. The average power removed from the wave per unit volume by means of absorption and scattering is

$$\frac{\omega}{2\pi} \int_0^{2\pi/\omega} J \cdot E \, dt = \frac{\omega}{2} X'' E_0^2 \qquad \left(J = \frac{dP}{dt}\right). \tag{1-70}$$

Equation (1–70) shows that the out-of-phase component of the electric susceptibility determines how much attenuation takes place as a plane wave traverses a medium. The time-averaged Poynting vector for the plane electromagnetic wave is given by the expression

$$\frac{c}{4\pi} \mathbf{E} \times \mathbf{B} = \frac{c}{8\pi} E_0^2 \hat{k} \tag{1–71}$$

where $\hat{k}$ is a unit vector in the direction of propagation of the wave. Equations (1–70) and (1–71) imply that the fraction of the power attenuated in a path length dL is

$$\frac{4\pi\omega}{c} X''(\omega)\, dL. \tag{1–72}$$

Although not much is known about the structure and composition of interstellar grains, the Kramers-Kronig dispersion relations, which relate the real and imaginary parts of the electric (or magnetic) susceptibility, make it possible to place some general limits on the amount of matter in grains and also on their equilibrium temperatures. The relevant Kramers-Kronig relation is

$$X'(0) = \frac{2}{\pi} \int_0^\infty \frac{X''(\omega)\, d\omega}{\omega}. \tag{1–73}$$

For spherical grains it can be shown that

$$X'(0) = n_g r^3 \left(\frac{\varepsilon_g - 1}{\varepsilon_g + 2} \right) \tag{1–74}$$

where r is the radius of the grain, and ε_g the dielectric constant of the grain material. If the density of grain material is assumed known ($\rho_g \sim 1$ g-cm^{-3}) and ε_g is taken to be about 2 or 3, then the measured interstellar extinction ($\simeq 0.5$ magnitude per kiloparsec at visual wavelengths), which is known as a function of wavelength in the visual and near infrared, can be used in conjunction with Equations (1–73) and (1–74) to place a lower limit on the amount of material in grains. This limit, which depends somewhat on the shape and dielectric constant of the grains, is about 10^{-27} g/cm^3.

If collisions with the surrounding gas can be neglected, the temperature equilibrium of the grains is the result of a balance between emission in the infrared and absorption of background radiation, which is mostly starlight unless the grains are shielded in dense clouds. The mean energy density of starlight in the interstellar medium is equivalent to a 3°K blackbody. However, much higher grain equilibrium temperatures are predicted (10–20°K) because grains are small as compared to the wavelengths of the emitted radiation and therefore inefficient radiators. It might be assumed that absorption features in the infrared might make it possible for the grains to radiate very effectively at these wavelengths. However, the Kramers-Kronig relation can be used to place upper limits on the amount of emission. These limits indicate that unless grains are in dense clouds, their equilibrium temperatures are likely to be at least $10°K$.

1-9 INTERSTELLAR MOLECULES

At the present time, many molecules are known to exist in interstellar space. The first molecules to be discovered were observed as absorption lines in stellar spectra. These molecules included CH, CH^+, and CN. Absorption and emission lines of OH were subsequently discovered at 18-cm wavelength. The very high brightness temperatures measured for the emission lines of OH are much too high to represent a true thermal temperature, and therefore it is plausible to assume that maser action is responsible for the emission. Recent measurements carried out by means of rocket-borne experiments have shown that the H_2 molecule is present in interstellar space. In addition, microwave absorption and emission lines have been discovered for a large number of interstellar molecules (for example, NH_3, H_2O, H_2CO, CO, CS, SiO, OCS, HCN, CH_3CN, HCOOH, HC_3N, and NHCO). The measured abundances and excitations of these molecules provide us with a great deal of information about physical conditions inside interstellar clouds. Moreover, it is remarkable that a number of complex organic molecules are present in interstellar space. Although interstellar molecules may not survive the formation of the solar system, their presence in interstellar space demonstrates that complex molecules are formed outside the laboratory, and therefore the formation of such molecules in planetary atmospheres is made more plausible. The connection with biology is particularly interesting because H_2O, NH_3, H_2CO, and HCN are starting points for primitive earth synthesis experiments whose products include NHCO and HC_3N.

Since the excess kinetic energy of colliding atoms must be carried away by the simultaneous emission of photons, most binary collisions between atoms or molecules will not lead to molecular formation. Although it is plausible that such molecules as CH, CH^+, and CN may be formed as a result of collisions in interstellar space, it is likely that many interstellar molecules (for example, H_2, OH, NH_3, and H_2O) are formed on the surfaces of grains where their excess energies of formation can be readily absorbed by adjacent atoms or the lattice of the grain.

An atom that is absorbed onto the surface of a grain will remain there for a time interval

$$\tau \simeq v_0^{-1} \exp\left[\frac{V_0}{kT_g}\right] \tag{1-75}$$

where $v_0 \simeq 10^{12}$–10^{13} sec^{-1} is the oscillation frequency of the atom trapped on the surface; T_g is the temperature of the grain; and V_0 is the depth of the surface potential well minus the quantum mechanical zero point energy. The reciprocal of the exponential term in Equation (1–75) is the probability that the atom will escape from the surface of the grain during a single oscillation. During its residence on a grain, an atom will diffuse over the surface. If H_2 molecules are to be formed at a significant rate the mean time between successive arrivals of H atoms,

$$\tau_c \simeq \left[\pi r_g{}^2 \left(\frac{kT}{m_H}\right)^{1/2} n_H \right]^{-1} \qquad (1\text{–}76)$$

must be shorter than the mean residence time τ for a H atom. In Equation (1–76), $r_g \simeq 10^{-5}$ cm is the radius of the grain; $T \simeq 20 - 100°K$ is the temperature of the gas; n_H is the number density of hydrogen atoms; and $(kT/m_H)^{1/2}$ is the mean velocity of hydrogen atoms. If $n_H = 10^2$ and $V_0/k = 200°K$, the critical grain temperature for the formation of H_2 molecules predicted by the condition $\tau_c < \tau$ is about $6°K$.

Diatomic Molecules

Diatomic molecules are the simplest molecules and also the first to be discovered in interstellar space. Because nuclei are much heavier than electrons, they move much more slowly. Therefore, in determining the electronic energy levels, we may assume that the nuclei are separated by a fixed distance R and that the electrons are moving in the potential of these nuclei. The electronic energy levels depend on R and are labeled $E_n(R)$. The problem of determining the molecular energy levels is thus separated into the distinct problems of finding the electronic levels and also determining the motions of the nuclei. A qualitative understanding of how the electronic energy levels depend on R can be seen from Figure 1–8. A molecule is in an equilibrium state when the nuclei are at a fixed distance R_e. The energy $E_n(R)$ is always positive for sufficiently small R since the Coulomb interaction of the two nuclei is positive. For R sufficiently large, the energy $E_n(R)$ must approach that of two separated atoms. The energy difference $E(R_e) - E(\infty)$ is called the dissociation energy of the molecule.

An electron in an atom moves in a field of spherical symmetry, and consequently all three components of its angular momentum remain constant in time. For this reason, two quantum numbers, the total angular momentum J and the z component of the angular momentum J_z, characterize its atomic state. On the other hand, a diatomic molecule possesses a field of axial rather than spherical symmetry, and consequently only the angular momentum about the axis of symmetry of the molecule remains constant as a function of time. To first approximation, the molecular states have a two-fold degeneracy (unless the angular momentum is zero) since the electronic energy level should not depend strongly on the sense of rotation (that is, clockwise or counterclockwise) about the molecular axis. Molecular states with angular

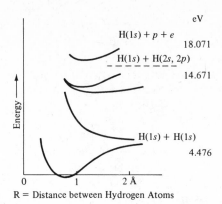

Figure 1-8   Some potential energy curves of the hydrogen molecule.

momenta $0, 1, 2, \ldots$ are called $\Sigma, \Pi, \Delta, \ldots$, respectively. Since the spin of the electrons has little interaction with orbital motions, molecular states are denoted $^{2S+1}\Sigma$, $^{2S+1}\Pi$, and so on, where S is the electron spin of the molecule.

If the center of mass of the two nuclei (m_1 and m_2) is chosen as the origin of the frame of reference, the kinetic energy of the nuclei can be written as

$$T = \tfrac{1}{2}m_r v^2 \qquad (1\text{-}77)$$

where

$$m_r = \frac{m_1 m_2}{m_1 + m_2}$$

is the reduced mass, and v the relative velocity of the two nuclei. The constant total energy H of the molecule becomes

$$H = \tfrac{1}{2}m_r v^2 + E_n(R) \qquad (1\text{-}78)$$

which after some manipulation can be rewritten

$$H = \tfrac{1}{2}m_r \left(\frac{dR}{dt}\right)^2 + E_n(R) + \frac{\mathbf{J}^2}{2m_r R^2} \qquad (1\text{-}79)$$

where $\mathbf{J}$ is the angular momentum of the two nuclei about their center of mass. The first two terms in the above equation represent the vibrational energy H_{vib}, and the last term represents the rotational energy H_{rot}.

If the amplitude of the vibrations about the equilibrium state are sufficiently small, the vibrational energy is

$$H_{\text{vib}} \approx (n + \tfrac{1}{2})\hbar\omega \qquad n = 1, 2, 3, \ldots. \qquad (1\text{-}80)$$

Since the rotation of the nuclei about their center of mass does not depend on the orientation of the molecule in space, we have

$$\mathbf{J}^2 = \hbar^2 J(J + 1) \qquad J = 0, 1, 2, \ldots. \qquad (1\text{-}81)$$

If R is approximately constant, the rotational energy becomes

$$H_{rot} \approx \frac{\hbar^2}{2I} J(J + 1) \qquad (1\text{–}82)$$

where I, the moment of inertia, is a constant. The energy differences between the low-lying rotational levels are, in general, much less than the energy differences between the vibrational levels.

H_2 Molecules

Recent UV measurements have confirmed the existence of the H_2 molecule in interstellar space. By measuring the excitation of molecules such as CO and HCN, the densities of interstellar clouds can be inferred. The apparent deficiency of 21-cm line radiation from certain dense clouds provides indirect evidence for the presence of H_2 molecules.

Although the dissociation energy of H_2 is only 4.48 eV, transitions at this energy are strongly forbidden since the H_2 molecule is homonuclear and therefore without an electric dipole moment. Photon energies in excess of 14.7 eV are required for direct photodissociation, while photons with energies in excess of 15.4 eV are required for photoionization. Since the UV radiation shortward of the Lyman limit (> 13.6 eV) is likely to be absent in HI regions, it was once believed that H_2 molecules might pervade interstellar space and even constitute a sizable fraction of the mass of the galaxy. However, it was subsequently pointed out that H_2 molecules can be destroyed by a dissociation into the vibrational continuum of the ground electronic states, which will occur with a significant probability following an excitation into a higher electron state through the absorption of a UV photon ($\lambda \lesssim 10^3$Å). If the H_2 molecule is in an excited vibrational level ($v \geq 2$) of the ground electron state, destruction of the H_2 molecule can also arise as a result of direct photoionization or photodissociation. This circumstance implies that H_2 molecules are likely to exist only in dense HI clouds where they can be produced relatively rapidly and shielded from interstellar UV radiation.

Because the H_2 molecule has no electric dipole moment, its formation is likely to arise on the surface of grains rather than by means of two-body collisions. Three-body collisions are infrequent in interstellar space because the density is very low. When two H atoms interact, they spend only about 10^{-13} sec within the range of their chemical forces. If molecular formation is to occur, a radiative transition from an upper state to a lower state must take place during the period of the interaction. Such a transition probability is, however, highly unlikely since the transition probability is very small (approximately 1 sec^{-1}) due to the absence of an electric dipole moment. The binary collision $H^- + H \rightarrow H_2 + e^-$ is likely to be more favorable for the formation of H_2 molecules than collisions between neutral H atoms.

OH Molecules

The degeneracy of the ground state of the OH molecules is split by an interaction between the rotation of the nuclei and the motion of the unpaired electron in its orbit. This phenomenon, which is called lambda doubling, is responsible for the observed 18-cm OH emission and absorption lines (see Figures 1–9 and 1–10).

Two different states of electronic motion are possible: The unpaired electron distribution may lie either along the axis of rotation or in the plane of rotation. The hyperfine interaction caused by the spin of the proton causes each level of the Λ doublet to be split into two states in which the spin of the proton ($S = \frac{1}{2}$) and the magnetic moment of the molecule ($J = \frac{3}{2}$ for ground

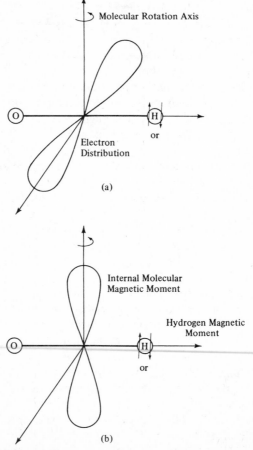

Figure 1–9 The two orientations of the electron distribution that cause Λ-doubling in the OH molecule.

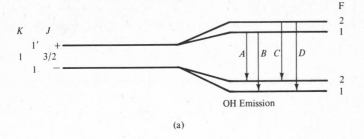

(a)

Transition	F Number Upper	F Number Lower	Frequency (MHz)	Relative Intensity
A	1	2	1612.231	1
B	1	1	1665.401	5
C	2	2	1667.358	9
D	2	1	1720.533	1

(b)

Figure 1–10   The lowest energy levels of the OH molecule. The relative intensities of the transitions are those that would be predicted if the levels were populated according to a Boltzmann distribution.

state) are either parallel ($\mathbf{F} = \mathbf{J} + \mathbf{S} = 2$) or antiparallel ($F = 1$) to each other.

There exists a number of classes of OH sources. These OH sources are distinguished by the anomalous intensity ratios of their emission lines. In one class of OH source found on the edge of H_2 regions, the 1665- and 1667-GHz transitions are the strongest. Another class of sources is associated with infrared objects and is most intense at 1612 GHz.

The OH molecule is also observed in absorption. Although its interstellar abundance is much less ($\sim 10^{-7}$) than that of hydrogen, the OH molecule has an electric dipole moment, and therefore the absorption cross section is much higher (by factor of about 10^4) than for the hydrogen atom. This circumstance explains why significant absorption due to the OH molecule is observed in the plane of the galaxy.

Interstellar Masers

The brightness temperatures measured for a number of OH and H_2O transitions exceed $10^{13} {}^\circ K$. It is highly implausible that such high temperatures are caused by thermal sources, and consequently it is generally believed that they are the result of maser amplification. The basic requirement for such amplification is that the relative populations of two (or more) energy levels separated by energy hv be inverted as compared to the usual Boltzmann distribution,

$$n_2 = n_1 e^{-(hv/kT)} \tag{1-83}$$

where n_1 is the lower energy state. For maser amplification, it is necessary that n_2 exceed n_1.

If k_v is the fractional absorption per unit distance at frequency, then the integrated absorption coefficient

$$\gamma = \int k_v \, dv. \tag{1-84}$$

It can be shown that

$$\gamma = \frac{8\pi^3 |\mu_{12}|^2 v}{3hc} (n_1 - n_2) \tag{1-85}$$

where μ_{12} is the electric dipole moment between levels 1 and 2. If $hv/kT \ll 1$, then we have

$$n_1 - n_2 = n_1 \frac{hv}{kT} \tag{1-86}$$

where n_1 is the number density in the lower state. From Equation (1-85) it follows that *negative* absorption, which implies amplification, will result if n_2 exceeds n_1.

The amount of amplification can be estimated by the following considerations. For an assumed linewidth Δv, the fractional change per unit length of the radiation is

$$\frac{1}{N_{12}} \frac{dN_{12}}{dl} = - \frac{\gamma}{\Delta v} \tag{1-87}$$

where $\gamma/\Delta v$ is negative for amplification, and N_{12} is the number of photons $cm^{-2} sec^{-1}$. Integrating Equation (1-87) between $l = 0$ and $l = L$, we find

$$N_{12} (l = L) = N_{12} (l = 0) e^{-(\gamma/\Delta v)L}. \tag{1-88}$$

Equation (1-88) suggests that the maser amplification can increase indefinitely. However, the onset of saturation will limit the amplification to some finite amount. Let P_{21} (and P_{12}) be the probabilities per second that the state of the molecules will be transferred from state 2 to 1 (and 1 to 2), respectively, without the emission or absorption of a photon at frequency v_{12}. The rate equation between the levels is

$$n_1 P_{12} = - \frac{\gamma}{\Delta v} N_{12} + n_2 P_{21}. \tag{1-89}$$

If P_{21} and P_{12} are constant, Equations (1-85), (1-88), and (1-89) imply that as L becomes sufficiently large, $-(\gamma/\Delta v)$ must become small, and n_1 must approach n_2.

Formaldehyde

Formaldehyde (H_2CO) is known to be present in relatively dense clouds throughout the galaxy. Some of these dense clouds are likely to be protostars. The discovery of interstellar H_2CO is particularly interesting because it is observed in absorption against the $3°K$ background radiation. This circumstance means that the temperatures as inferred by the relative populations of the energy levels ($n_2/n_1 = e^{-h\nu/kT}$) are less than $3°K$. This result may seem surprising in view of the presence of a universal background radiation. However, just as it is possible for molecular excitation to overpopulate upper levels and thereby produce maser action, it is possible to overpopulate lower levels and produce a level population that is less than $3°K$.

1–10 DIFFUSE NEBULAE (HII REGIONS)

Large numbers of extended emission regions are observed at low galactic latitudes. These ionized diffuse nebulae, which constitute about 5 percent of the interstellar gas, are produced by neighboring hot stars. Visual emission lines such as hydrogen recombination lines (Hα, Hβ, Hγ) and collisionally excited forbidden lines (for example, [OII], [OIII], and [NII]) are also observed. The temperatures and densities inside HII regions can be estimated by measuring the ratios of intensities of appropriately chosen emission lines. The kinetic temperatures of HII regions are typically $10^{4}°K$, and the electron densities vary from $n_e \sim 5–10^3$ cm^{-3}.

When a hot star turns on close to an extensive HI region, it will ionize the neutral hydrogen in its immediate vicinity. Initially, the ionization front will move away from the star at supersonic velocities. However, the gas will not move significantly. The motion of the ionization front will be slowed when its radius approaches the initial "Strömgren" radius R_i at which point the number of ionizations will equal the number of recombinations inside the ionized sphere. For a sphere of constant density, the condition that N_{L_c}, the number of emitted photons per second beyond the Lyman limit, equal the recombination rate is

$$N_{L_c} = \frac{4\pi}{3} n_e^2 \alpha(T) R_i^3 \qquad (1\text{–}90)$$

where $\alpha(T)$ is the hydrogen recombination coefficient, n_e the number density of electrons, and R_i the radius of the ionized region. Equation (1–90) implies that

$$R_i = \left[\frac{3}{4\pi} \frac{N_{L_c}}{n_e^2 \alpha(T)} \right]^{1/3}. \qquad (1\text{–}91)$$

When the ionization front has reached the radius R_i, a pressure imbalance will exist between the un-ionized gas and the ionized gas since the temperature inside the ionized sphere is much higher than in the un-ionized gas, but the densities are approximately the same. The existence of a pressure imbalance means that the ionized region will continue to expand until its density is sufficiently low for the condition $n_1 T_1 \simeq n_2 T_2$ to hold across the front. It can readily be shown that the radius and mass of the final ionized sphere are

$$R = \left(\frac{2T_2}{T_1}\right)^{2/3} R_i$$

and   (1–92)

$$M = \left(\frac{2T_2}{T_1}\right) M_i$$

where T_2 and T_1 are the temperatures of the ionized and un-ionized regions, respectively, and the emitted radiation from the star has been assumed constant as a function of time.

We have implicitly assumed that the boundary between the ionized and neutral regions is much smaller than the regions themselves. The width of the boundary between neutral and ionized gas is approximately the mean free path of an ionizing photon through neutral gas. For UV radiation, this path length is

$$\lambda \simeq \frac{1}{n_H \sigma} \simeq \frac{0.05}{n_H} \text{ pc}$$ (1–93)

which is much less than the diameters of HII regions.

Because of their large cross sections for absorption, resonance radiation such as Lyman α is absorbed and reemitted many times before it can escape from HII regions. It has been estimated that the Lyman α optical depth is about $\tau_\alpha = 10^5$ for a typical HII region. It is plausible that a Lyman α photon will be absorbed as a consequence of some relatively infrequent process before it has escaped from the nebula. Two such absorption processes are absorption by dust particles and conversion into two photon continuum photons. The latter process, which is caused by collisions of excited hydrogen atoms with electrons, protons, and ions, is described by means of the reactions:

$$H(1S) + L\alpha \rightarrow H(2P)$$
$$H(2P) + e^- \rightarrow H(2S) + e^-$$

or

$$H(2P) + p \rightarrow H(2S) + p$$
$$H(2S) \rightarrow H(1S) + 2 \text{ continuum photons.}$$

The above process takes place with a probability of

$$\frac{C_{2P,2S}}{A_{2P,1S} + C_{2P,2S}} \simeq \frac{C_{2P,2S}}{A_{2P,1S}} \simeq 2 \times 10^{-13} n_e$$ (1–94)

where $C_{2P,2S}$ is the probability per unit time that the $2P$ state of hydrogen will be collisionally changed into a $2S$ state, and $A_{2P,1S}$ is the probability for radiative decay.

If the scattering of a Lyman α photon were completely coherent, the number of scatterings necessary for a photon to escape an HII region would be

$$\frac{t_D}{c\lambda} = \frac{(R^2/c\lambda)}{c\lambda} = \tau_\alpha^2 \qquad (1-95)$$

where t_D is the diffusion time; R is the radius of the HII region; and λ the mean free path for scattering. However, an absorbed photon will not be reradiated coherently but rather with a frequency distribution that reflects the Doppler distribution of the atoms. This circumstance implies that the probability that the frequency of the reemitted photon is between Δv, and $\Delta v + d\,\Delta v$ about the rest frequency v_0 is

$$P\left(\frac{\Delta v}{\Delta v_D}\right) = \frac{1}{\pi^{1/2}} \exp\left[-\left(\frac{\Delta v}{\Delta v_D}\right)^2\right] \qquad (1-96)$$

where

$$\Delta v_D = \frac{v_0}{c}\left(\frac{2kT}{m_H}\right)^{1/2}$$

is the Doppler width; m_H is the mass of a hydrogen atom; and T is the temperature of the gas. The redistribution in frequency that arises after each scattering implies a greatly reduced number of scatterings since the few photons emitted in the wings of the line will be able to escape directly from the nebula. For this reason, the probability for 2-photon continuum emission is likely to be small in HII regions although it may become appreciable in some planetary nebulae.

Radio-frequency transitions between energy levels of very high quantum number ($n \gtrsim 100$) are observed in HII regions. The strongest lines are those of H and He. Because states of high quantum number are populated primarily by means of electronic recombinations, they are called radio-frequency recombination lines. For a hydrogenic atom, the frequency of transition from an upper state n to a lower state m is

$$v = cRZ^2(m^{-2} - n^{-2}) \simeq 2cRZ^2m^{-3}(n - m) \qquad (1-97)$$

where R is the Rydberg constant; Z is the charge of the nucleus; and the assumption $(n - m)/n \ll 1$ has been used in writing down the right-hand expression in Equation (1–97). Atoms that have electrons in levels with very high n have very large radii. Transitions between levels of high quantum number can be detected in interstellar space because the line width of the transition is caused by Doppler broadening, which is proportional to the frequency of the transition and also because the very low collision rate in interstellar space implies that the intrinsic widths of the energy levels are determined by the radiative lifetimes and not the collision rate as is the case in the laboratory.

Since departures from local thermodynamic equilibrium are expected to be the same for H and He^+, the relative strengths of H and He^+ lines can be used to determine relative abundances of He with respect to H in HII regions. These measurements give $N(He)/N(H) \simeq 0.08 - 0.09$. Radio-frequency recombination lines can be used to map out the internal velocities of HII regions and also to trace out spiral arm patterns.

1–11 GALACTIC ROTATION

The rotation of the galaxy can be studied by means of both optical and radio measurements. Optical measurements, which are usually limited to distances <4 kpc from the sun because of extinction by interstellar dust particles, include the determination of stellar motions (especially those of RR Lyrae stars) and observations of regions of ionized hydrogen called HII regions. On the other hand, 21-cm line measurements can be used to study the motion of the inner regions of the galaxy because interstellar dust particles have dimensions ($\sim 10^{-5}$ cm) much less than radio wavelengths and therefore do not appreciably absorb or scatter 21-cm radiation. More recently radio recombination lines have been used to study the properties of HII regions.

We assume that the motion of the galactic disk is one of circular differential rotation about the axis of galactic coordinates. Let R be defined as the distance from the axis. We assume that the angular velocity $\omega = \omega(R)$ is a monotonically decreasing function of R. The motion of the local standard of reference and an arbitrary star s with respect to the galactic center is shown in Figure 1–11. If v_0 and v_s are the magnitudes of the velocity of the local rest frame and an arbitrary star s, respectively, the radial velocity of s with respect to 0 is

$$\begin{aligned} v_{rad} &= v_s \cos \alpha - v_0 \sin l \\ &= R_0(\omega - \omega_0) \sin l \end{aligned} \tag{1-98}$$

with

$$\begin{aligned} v_s &= \omega R \\ v_0 &= \omega_0 R_0. \end{aligned}$$

The last step follows from the law of sines. For stars that are at small distances from the local standard of reference, v_{rad} can be rewritten as

$$v_{rad} = -2A(R - R_0) \sin l \tag{1-99}$$

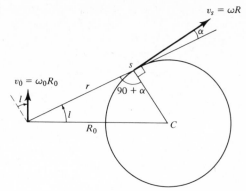

Figure 1-11   The motion of the local standard of rest and an arbitrary star s with respect to the galactic center, which is denoted by C.

where we have used the relation

$$\omega - \omega_0 = \left(\frac{d\omega}{dR}\right)_{R_0} (R - R_0)$$

and A is defined as

$$A \equiv -\frac{1}{2} R_0 \left(\frac{d\omega}{dR}\right)_{R_0} = \frac{1}{2}\left(\frac{v_0}{R_0} - \left(\frac{dv}{dR}\right)_{R_0}\right).$$

If the distance r is sufficiently small (that is, $r \ll R_0$)

$$r \simeq -\frac{(R - R_0)}{\cos l} \tag{1-100}$$

and it follows from Equations (1–99) and (1–100) that

$$v_{\text{rad}} \simeq rA \sin 2l. \tag{1-101}$$

Equation (1–101) shows that there exist two maxima and minima as l goes from 0–360°.

The proper motion of s with respect to the local standard of rest is the tangential velocity v_{tan} divided by the stellar distance, that is,

$$\frac{v_{\text{tan}}}{r} = \frac{\omega R}{r} \sin \alpha - \frac{\omega_0 R_0}{r} \cos l \tag{1-102}$$

where

$$B = A - \frac{v_0}{R_0} = -\frac{1}{2}\left(\frac{v_0}{R_0} + \left(\frac{dv}{dR}\right)_{R_0}\right).$$

The quantities A and B are called the Oort constants. The values for the Oort constants obtained from the study of stellar motions are

$$A = 15 \text{ km sec}^{-1} \text{ kpc}^{-1}$$
$$B = -10 \text{ km sec}^{-1} \text{ kpc}^{-1}. \tag{1-103}$$

The distance to the galactic center R_0 can be obtained by determining the centroid of globular clusters and the absolute visual magnitude for RR Lyrae stars. If we assume $M_v = 0.5$ for RR Lyrae stars, then R_0 is approximately 8–10 kpc and $v_0 \sim 250$ km/sec.

We can determine the maximum radial velocity along the line of sight in any direction by means of the observed 21-cm line profile. From Equation (1–99) it follows that v_{rad} is a maximum for a given galactic longitude l if R is a minimum. It can be seen from Figure 1–11 that

$$R_{min} = |\sin l| R_0 \tag{1-104}$$

and therefore

$$v_{rad} (\max) = 2AR_0(1 - |\sin l|) \sin l. \tag{1-105}$$

The above equation shows that AR_0 can be found from 21-cm line measurements. Since A can be determined from optical measurements, we can estimate R_0. This independently derived value of R_0 is in approximate agreement with our above-mentioned value.

1–12 GALACTIC MAGNETIC FIELD

In addition to gas and dust, the interstellar medium is known to be pervaded by a large scale magnetic field. Direct evidence for an interstellar magnetic field can be obtained by means of Faraday rotation measurements of the plane of polarization of linearly polarized radiation from discrete extragalactic radio sources and also from the Zeeman splitting of the 21-cm neutral hydrogen line. The existence of a galactic magnetic field may also be inferred from the physical properties of the galactic background radio emission and the measured polarization of starlight.

The Lorentz force on a free electron is

$$m\ddot{\mathbf{r}} = -e\mathbf{E} - \frac{e}{c} \dot{\mathbf{r}} \times \mathbf{B}. \tag{1-106}$$

It is assumed that the magnetic field $\mathbf{B}$ is uniform; the electric field $\mathbf{E}$ is due to an incident electromagnetic wave of frequency ω; and the magnitude of the oscillating component of the magnetic field is small as compared to its static component. If the z axis is chosen along the magnetic field **B**, then the x and y components of Equation (1–106) become

$$m\ddot{x} = -eE_x - \frac{e}{c}\dot{y}B \qquad (1\text{–}107a)$$

$$m\ddot{y} = -eE_y + \frac{e}{c}\dot{x}B. \qquad (1\text{–}107b)$$

The oscillating electric field is the superposition of right-handed (RH) and left-handed (LH) circularly polarized components, that is,

$$E_{\pm} = E_x \pm iE_y = E_0 e^{\pm i\omega t} \qquad (1\text{–}108)$$

where the plus sign and minus sign denote right-handed and left-handed polarization, respectively. Multiplying Equation (1–107b) by i and then adding Equations (1–107a) and (1–107b), we find

$$\ddot{\xi}_{+} \equiv \ddot{x} + i\ddot{y} = -\frac{e}{m}(E_x + iE_y) + \frac{ie}{mc}B(\dot{x} + i\dot{y}). \qquad (1\text{–}109)$$

A similar expression can be found if we multiply Equation (1–107b) by $-i$, that is,

$$\ddot{\xi}_{-} \equiv \ddot{x} - i\ddot{y} = -\frac{e}{m}(E_x - iE_y) - \frac{ie}{mc}B(\dot{x} - i\dot{y}). \qquad (1\text{–}110)$$

The solutions to Equations (1–109) and (1–110) represent the displacements of the charge caused by right-handed and left-handed circularly polarized electromagnetic waves, respectively. Making the substitutions $\xi_{\pm} = a_{\pm}e^{\pm i\omega t}$ into Equations (1–109) and (1–110), the solutions for the displacement of the charge become

$$a_{\pm} = \frac{(eE_0/m\omega^2)}{[1 \mp (eB/mc\omega)]} \qquad (1\text{–}111)$$

where the displacements are out of phase with the force $-eE_0$. The corresponding expressions for the polarization, which is the electric dipole moment per unit volume, become

$$P_{\pm} = -n_e e a_{\pm} e^{\pm i\omega t} = \frac{-(n_e e^2 E_0 e^{\pm i\omega t}/m\omega^2)}{[1 \mp (eB/mc\omega)]}. \qquad (1\text{–}112)$$

Since the dielectric constant and polarization are related by the equation

$$\mathbf{D} = \varepsilon\mathbf{E} = \mathbf{E} + 4\pi\mathbf{P} \qquad (1\text{–}113)$$

the indices of refraction for right- and left-handed circularly polarized radiation become

$$n_\pm{}^2 = \varepsilon_\pm = 1 - \frac{4\pi n_e e^2/m\omega^2}{1 \mp (eB/mc\omega)} \tag{1-114}$$

from which the corresponding phase velocities $v_\pm = c/n_\pm$ can be evaluated. If

$$\omega \gg \omega_p = \sqrt{\frac{4\pi n_e e^2}{m}}$$

and

$$\omega \gg \frac{eB}{mc}$$

then $n_\pm{}^2$ can be rewritten as

$$n_\pm = 1 - \frac{1}{2}\frac{4\pi n_e e^2}{m\omega^2}\left(1 \pm \frac{eB}{mc\omega}\right). \tag{1-115}$$

Equation (1–115) implies that the Faraday rotation angle of the plane polarization over a path length L is

$$\psi = \frac{1}{2}\left(\frac{2\pi L}{\lambda_+} - \frac{2\pi L}{\lambda_-}\right) = \frac{\pi L \nu}{c}(n_+ - n_-) \tag{1-116}$$

where the final relation is correct if $n_\pm \simeq 1$. Since B and n_e are not necessarily uniform along the path of the radiation, a more general expression for ψ is

$$\psi \equiv (RM)\lambda^2 = \left(0.81 \times 10^6 \int_0^L n_e B \cos\phi \, ds\right)\lambda^2 \tag{1-117}$$

where the rotation measure (RM) is in units of radians per m^2; λ is in meters; n_e in cm^{-3}; B in gauss; and s and L are in units of parsecs. ϕ denotes the angle between $\mathbf{B}$ and $\mathbf{s}$.

A large fraction of extragalactic radio sources are linearly polarized at decimeter and shorter wavelengths. There is a strong correlation between the measured rotation measure and the galactic latitude of the source. The rotation measure is higher on the average for sources of low galactic latitude than for those of high galactic latitude and consequently indicates that the galactic magnetic field is concentrated toward the plane of the galaxy. By observing radio sources at several wavelengths, the position angle of the plane of polarization ψ can be measured as a function of λ^2. The measured λ^2 dependence of ψ as a function of galactic latitude proves the existence of a large scale galactic magnetic field and provides an estimate of $\langle n_e B L \rangle$ [see Equation (1–117)]. Measurements of the dispersion of radio pulses from pulsars make it possible to estimate n_e, and consequently crude values for the interstellar magnetic field can be obtained ($B \sim 2 \times 10^{-6}$ G).

In the presence of a weak magnetic field **B**, the $F = 1$ hyperfine state of neutral hydrogen will be split by means of the Zeeman effect into three levels (see Figure 1–12) whose energies with reference to the ground $F = 0$ state are

$$E = E_0 + \frac{e\hbar}{2mc} B$$

$$= E_0 \qquad\qquad (1\text{–}118)$$

$$= E_0 - \frac{e\hbar}{2mc} B$$

where $E_0 = h v_0$ is the undisturbed energy difference between the $F = 1$ and $F = 0$ hyperfine states. If we observe 21-cm radiation along the direction of the magnetic field, then the undisturbed 21-cm line will be split into two

Proton Electron

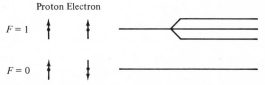

$F = 1$

$F = 0$

Figure 1–12   The hyperfine levels of the ground state of the hydrogen atom.

absorption lines that are centered at $v_0 \pm eB/4\pi mc$, respectively. One of these lines will absorb right-handed circularly polarized radiation; the other will absorb left-handed circularly polarized radiation. For this reason, if we measure the absorption of right-handed and left-handed circularly polarized radiation separately as a function of frequency, we can infer the magnetic field strength from the expression

$$v_{RH} - v_{LH} = \frac{eB}{2\pi mc} \qquad (1\text{–}119)$$

where v_{RH} (v_{LH}) is the frequency at which there is maximum absorption of right-handed (left-handed) circularly polarized radiation. Since the actual magnetic fields in interstellar space are not uniform, the Zeeman splitting will broaden the observed absorption lines. Most attempts to detect the Zeeman splitting have yielded negative results. These experiments indicate that the magnetic field in interstellar HI regions is typically $\leq 5 \times 10^{-6}$ G. However, magnetic fields in excess of 10^{-5} G have been measured in isolated interstellar clouds.

As will be discussed in Section 1–13, there is direct evidence for the existence of very-high-energy electrons in interstellar space. If such electrons move in a magnetic field, they will radiate by means of magnetobremmstrah-lung (also called synchrotron radiation—see Section 9–3) at frequencies up to about

$$v_{max} \simeq \frac{eB}{2\pi mc} \left(\frac{E}{mc^2} \right)^2 \tag{1–120}$$

where E and m are the electron's energy and mass, respectively. The intensity, approximate power law spectrum, and linear polarization of the measured galactic background radiation is consistent with the assumption that at least for frequencies $\leq 10^9$ Hz, it is caused by magnetobremsstrahlung of relativistic electrons moving in magnetic fields of $\sim 2 \times 10^{-6}$ G.

The observed polarization of starlight provides important evidence concerning galactic magnetic fields. Because reddened stars are preferen-tially polarized and the plane of polarization shows a strong preference for the galactic plane, the polarization of starlight is undoubtedly an interstellar phenomenon that is not primarily associated with individual stars. It is highly likely that stellar polarization is the result of absorption by means of elon-gated dust particles that have been preferentially aligned along the galactic magnetic field. The Davis-Greenstein theory of grain alignment assumes that grains are kept spinning by collisions with interstellar gas and that they tend to become aligned as a consequence of dissipation caused by interaction with the interstellar magnetic field. Preferentially aligned grains will scatter or absorb more of the component of the radiation whose electric field vector is parallel to the long axis of the grain and therefore normal to the magnetic field. This circumstance means that the plane of polarization of the observed radiation is predicted to be in the plane of the **B** field.

The magnetic alignment mechanism assumes that grains contain para-magnetic impurities (that is, impurities with permanent magnetic dipole moments) and that paramagnetic relaxation couples the orientation of grains to the interstellar magnetic field. Such coupling produces an isotropy in the distribution of angular momentum because although rotation is damped in two space axes, it is not damped about the direction of the magnetic field. The rate of loss of angular momentum of the grain rotating with angular velocity ω is

$$\frac{dJ}{dt} = \frac{1}{\omega} \frac{dE}{dt} = \frac{V}{\omega} \left(\frac{\omega}{2\pi} \right) \int_0^{2\pi/\omega} \mathbf{B} \cdot \frac{d\mathbf{M}}{dt} dt \tag{1–121}$$
$$\simeq X'' B^2 V$$

where **M** is the magnetization, X'' the imaginary part of the magnetic sus-ceptibility, E the kinetic energy of the grain, and V the volume of the grain. From Equation (1–121) we can define a characteristic magnetic damping time. We find

$$\tau_m \simeq \frac{J}{X'' B^2 V} = \frac{I\omega}{X'' B^2 V} \tag{1–122}$$

where I is the moment of inertia of the grain.

Collisions with the surrounding interstellar gas and cosmic ray particles impart random amounts of angular momentum to the rotating grains. An atom that collides with a grain is likely to stick to its surface at least temporarily and then evaporates after diffusing to a different position where it is ejected with an energy characteristic of the surface temperature. The time scale for the damping of rotational energy by means of collisions is

$$\tau_R \simeq \frac{M}{\rho \langle v \rangle A} \qquad (1\text{-}123)$$

where M is the mass of the grain; A is its effective area; ρ is the density of the gas; and $\langle v \rangle$ is the mean velocity of the gas particles. Since the degree of alignment depends on a number of factors including the assumed shape of the grain, detailed calculations are required to calculate it accurately. However, the critical parameter in determining the extent of the alignment is the ratio τ_m/τ_R.

Let us assume that a grain is a prolate spheroid. In order to achieve grain alignment, the symmetry axis must be correlated with its angular momentum $\mathbf{J}$. Such a correlation is likely to be achieved because, in equilibrium, the rotational kinetic energy will tend to be equal about all principal axes of the grain (that is, $\frac{1}{2}kT_\parallel \sim \frac{1}{2}kT_\perp$), and therefore the value of $J = I\omega$ will be least about the axis for which the moment of inertia is least.

Magnetic Flux Constancy

If σ is the electric conductivity and v the velocity of the gas with respect to the magnetic field, Ohm's law implies

$$\mathbf{J} = \sigma \mathbf{E}' = \sigma \left(\mathbf{E} + \frac{\mathbf{v}}{c} \times \mathbf{B} \right) \qquad (1\text{-}124)$$

where $\mathbf{E}'$ is the electric field in the frame of reference of the gas. From Equation (1-124) and the Maxwell equation

$$\nabla \times \mathbf{E} = -\frac{1}{c} \frac{\partial \mathbf{B}}{\partial t} \qquad (1\text{-}125)$$

it follows that

$$\nabla \times \left[\frac{\mathbf{v}}{c} \times \mathbf{B} - \frac{\mathbf{J}}{\sigma} \right] = \frac{1}{c} \frac{\partial \mathbf{B}}{\partial t} . \qquad (1\text{-}126)$$

If the accumulation of charge at localized regions of space can be neglected, the conservation of charge implies

$$\nabla \cdot \mathbf{J} = 0. \qquad (1\text{-}127)$$

Taking the curl of the Maxwell equation

$$\nabla \times \mathbf{B} = \frac{4\pi \mathbf{J}}{c} \qquad \text{(neglect displacement current)} \qquad (1\text{-}128)$$

implies that Equation (1–126) can be reduced to the equation

$$\mathbf{V} \times \left(\frac{\mathbf{v}}{c} \times \mathbf{B}\right) + \frac{c}{4\pi\sigma} \nabla^2 \mathbf{B} = \frac{1}{c} \frac{\partial \mathbf{B}}{\partial t} \tag{1–129}$$

where we have used the equation

$$\mathbf{V} \cdot \mathbf{B} = 0. \tag{1–130}$$

If the gas is at rest, then Equation (1–129) reduces to the diffusion equation

$$\frac{c}{4\pi\sigma} \nabla^2 \mathbf{B} = \frac{1}{c} \frac{\partial \mathbf{B}}{\partial t}. \tag{1–131}$$

It follows from dimensional analysis that the diffusion time scale for decay of the magnetic field is

$$\tau \sim \frac{4\pi\sigma}{c^2} L^2. \tag{1–132}$$

For conductors of small dimensions such as are found in the laboratory, this time scale is small. However, for ionized or partially ionized masses of gas with cosmic dimensions, τ is very long and therefore Equation (1–129) becomes

$$\mathbf{V} \times (\mathbf{v} \times \mathbf{B}) = \frac{\partial \mathbf{B}}{\partial t}. \tag{1–133}$$

Let C be a small closed contour that is frozen into a fluid moving with velocity v through a magnetic field. The total rate of change of the magnetic flux Φ is the result of two contributions. The first of these is the result of the explicit time dependence of the magnetic field. We have

$$\frac{\partial \Phi}{\partial t} = \int \frac{\partial \mathbf{B}}{\partial t} \cdot d\mathbf{A} \tag{1–134}$$

where the integration is carried out over a surface bounded by the contour C. An additional contribution comes about because of the motion through the magnetic field of the surface bounded by C. The resultant change in magnetic flux is

$$\oint \mathbf{B} \cdot (\mathbf{v} \times d\mathbf{s}) = -\oint (\mathbf{v} \times \mathbf{B}) \cdot d\mathbf{s} = -\int \mathbf{V} \times (\mathbf{v} \times \mathbf{B}) \cdot d\mathbf{A} \tag{1–135}$$

where Stokes' theorem has been used to obtain the second equality. From Equation (1–133) it follows that the total rate of change of magnetic flux, which is the sum of Equations (1–134) and (1–135), is equal to zero. This result implies that the magnetic lines of force are constrained to move with the gas so long as the diffusion time scale given in Equation (1–132) is sufficiently long.

1–13 COSMIC RAYS

The existence of very-high-energy particles called cosmic rays proves that there exist natural accelerators far more effective than those that can be constructed on earth. Although the origin of cosmic rays is not fully understood, supernova explosions, pulsars, exploding galactic nuclei, radio galaxies, and quasistellar objects are the most likely sites for their production.

The energy density of cosmic rays in the vicinity of the sun is 10^{-12} erg/cm³, and consequently cosmic rays are one of the main constituents of interstellar space. Relativistic protons and heavier nuclei are the principal components of cosmic radiation. High-energy electrons and positrons constitute approximately 1 percent of the total number density of cosmic rays and are responsible for most of the galactic radio background radiation.

The measured cosmic ray flux contains several orders of magnitude more light nuclei such as Li, Be, B, and He^3 per proton than indicated for normal cosmic abundances. It is plausible to assume that these anomalously high abundances are not representative of cosmic radiation at its source but the result of spallation of heavier nuclei in the interstellar medium. If the above assumption is made, then it can be shown that cosmic rays have traversed approximately 3 g/cm² of interstellar matter. If cosmic rays are confined to the galactic plane where $n_H \simeq 0.5$ cm⁻³, then the inferred lifetime of cosmic ray particles in the galactic disk is $\tau \simeq 2 \times 10^6$ years. On the other hand, if the cosmic rays were confined to the galactic halo where $n_H \simeq 0.01$ cm⁻³, then $\tau \simeq 10^8$ years. The observed energy density and the inferred lifetimes of cosmic rays indicate that $\sim 10^{41}$ erg/sec is needed to maintain the present interstellar flux.

The energy spectrum of cosmic ray particles is described approximately by a power law, that is,

$$N(E) \, dE = KE^{-\gamma} \, dE \qquad (\gamma \sim 2.6). \qquad (1\text{–}136)$$

We observe that although very energetic ($E \gtrsim 10^{20}$ eV) particles are detected, the cosmic ray energy density is predominantly in the form of relatively low-energy ($\sim 10^{10}$ eV) particles. The change in curvature of the spectrum for energies $E \gtrsim 10^{16}$ eV suggests that very-high-energy particles may have a different origin than those of lower energy (see Figure 1–13).

Because they are deflected by means of galactic magnetic fields, cosmic ray particles do not traverse interstellar space in a straight trajectory. The equation of motion of a charged particle in a magnetic field is

$$\frac{d\mathbf{p}}{dt} = eZ \frac{\mathbf{v}}{c} \times \mathbf{B}. \qquad (1\text{–}137)$$

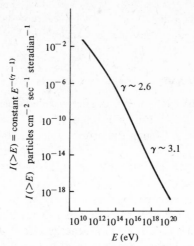

Figure 1-13   The number of cosmic ray particles cm^{-2} sec^{-1} steradian^{-1} with energy greater than E is shown as a function of E.

The motion of a charged particle in a uniform magnetic field is that of helical motion (see Figure 1–14). It follows from Equation (1–137) that the gyro-frequency of the particle is

$$\omega_B = \frac{eZB}{mc\gamma}, \qquad \gamma = \frac{1}{\sqrt{1 - (v^2/c^2)}}. \tag{1–138}$$

The radius of gyration of an ultrarelativistic particle in a magnetic field becomes

$$r_c = \frac{c \sin \theta}{\omega_B} = \frac{E \sin \theta}{eZB} \tag{1–139}$$

Figure 1-14   The helical motion of an electron in a magnetic field is shown schematically.

where E is the energy of the particle, and θ is the angle between the velocity of the particle and the magnetic field. Equation (1–139) shows that even protons whose energy is as high as 10^{17} eV have radii of curvature that are much less than the size of the galaxy.

In the presence of magnetic field inhomogeneities or external forces, the motion of a charged particle will not be pure helical motion, but will include a drift velocity across the magnetic lines of force. The direction of this drift velocity will be perpendicular to the magnetic field gradient or to the external force. If the characteristic scale of the magnetic field inhomogeneity is l, the drift velocity is approximately

$$v_D \sim \frac{r_c}{l} c \sin \theta = \frac{E \sin^2 \theta}{eZBl} \tag{1–140}$$

which is expected to be quite small for most cosmic ray particles in interstellar space.

Cosmic ray particles have an important influence on the interstellar medium. First, they are likely to be a principal source of heating and ionization in normal interstellar clouds. In addition, cosmic rays are likely to play an important dynamical role in interstellar space since the cosmic ray gas and the interstellar gas are coupled by means of the galactic magnetic field.

We assume that the interstellar medium, which is a composite fluid of thermal gas, dust, magnetic fields, and cosmic rays, is in hydrostatic equilibrium. The equation describing this equilibrium in the z direction can be written as

$$(1 + \alpha + \beta)\frac{1}{3\rho}\frac{d}{dz}(\rho\langle v^2 \rangle) = -g_z \tag{1–141}$$

where

$$P_{gas} = \tfrac{1}{3}\rho\langle v^2 \rangle = \rho\langle v_z{}^2 \rangle$$

cosmic ray pressure $= \alpha P_{gas}$

$$\frac{B^2}{8\pi} = \beta P_{gas} \qquad (\alpha \text{ and } \beta \text{ are assumed constant}).$$

If we assume that $\langle v^2 \rangle$ is a constant and that we can write

$$g_z = \left(\frac{dg_z}{dz}\right) z$$

with dg_z/dz a constant and z the distance from the galactic plane, the solution to Equation (1–141) becomes

$$\rho(z) = \rho(0)e^{-(z/h)^2} \tag{1–142}$$

with $h = 130(1 + \alpha + \beta)^{1/2}$ pc.

The effective thickness of the interstellar gas $2H$ is defined to be the total mass per square centimeter divided by $\rho(0)$, that is,

$$2H = \pi^{1/2}h = 230(1 + \alpha + \beta)^{1/2} \text{ pc.} \qquad (1\text{–}143)$$

The observed interstellar gas density together with the above expression for the effective thickness of the interstellar gas indicate that large scale interstellar magnetic fields $\gtrsim 5 \times 10^{-6}$ G are unlikely and that interstellar cosmic ray energy densities cannot greatly exceed those measured in the vicinity of the earth.

CHAPTER 2
Introduction to Stellar Theory

2-1 INTRODUCTION

Three basic time scales (dynamical, thermal, and nuclear) characterize the evolution of a star. Among these, the dynamical (or free-fall) time scale,

$$\tau_D \sim \left(\frac{R^3}{GM}\right)^{1/2}, \tag{2-1}$$

is generally the shortest. The dynamical time scale indicates how long it will take the displacement of an element to become large if the gravitational force on the element is not balanced by a pressure gradient. If a star is in quasi-hydrostatic equilibrium, then it follows from the virial theorem that the free-fall time is approximately twice the time for sound to traverse its radius (see Section 4-2). The thermal (or Kelvin) time scale is

$$\tau_K \sim \frac{GM^2}{RL} \tag{2-2}$$

where L, the luminosity, is the total energy emitted per unit time. The thermal time scale measures how long a star can maintain its luminosity without changing its radius appreciably if gravitational contraction is the only significant energy source. In discussing thermally unstable stars in Section 5-4, it will be useful to define a local thermal time scale

$$\tau_K \sim \frac{3P/2\rho}{\varepsilon}$$

where ε is the rate of nuclear energy generation in units of erg/g/sec.

As a star contracts, its central temperature increases. The main-sequence phase of stellar evolution begins with the ignition of hydrogen burning in the stellar core. Approximately 0.007 of the rest mass energy of a star can be converted into thermal energy during hydrogen burning since

$$(4m_p - m_{He})c^2 = 26.7 \text{ MeV}. \tag{2-3}$$

Hydrogen burning represents nearly 80 percent of the total nuclear energy content of a star. Because the nuclear energy content greatly exceeds that available by means of contraction (except during final gravitational collapse), the main-sequence lifetime

$$\tau_{ms} \simeq \frac{0.12E_n}{L} \tag{2-4}$$

where E_n is the total nuclear energy content of the star, and 0.12, the fraction of hydrogen burned before the star turns off the main sequence, is much longer than a thermal time scale. This circumstance means that if the effect of rotation and magnetic fields is small, the history of a star during and after the main-sequence phase should depend only on its mass and initial chemical composition.

Progress in the development of stellar theory has been greatly enhanced by the circumstance that stars are equilibrium configurations. Evidence for the long-term stability of the sun is provided by the continuing existence of life, which is dependent upon the stability of the sun over geological time scales. Additional evidence that many stars are equilibrium configurations follows from the remarkable constancy of the observed periodicities of pulsars and Cepheids.

Because there exists a rich store of observational information about stars, it is possible to test many of the predictions of stellar theory. The principal observable stellar parameters are stellar magnitude, distance, surface temperature, spectrum, and mass.

The apparent magnitude of an astronomical object is a measure of the amount of radiation within some specified bandwidth (for example, the standard visual (V), blue (B), or ultraviolet (U) filters) that is received at the earth. The magnitude scale is defined so that if we receive radiation in the amounts l_A and l_B from two objects, the difference between the magnitudes of these objects is

$$m_A - m_B = -2.5 \log\left(\frac{l_A}{l_B}\right) \tag{2-5}$$

and therefore magnitude differences indicate relative amounts of received radiation. This magnitude scale is not complete unless we define the magnitude of some standard object.

Stellar theory predicts the luminosity (that is, power output) from a star. For this reason, the apparent magnitude is not a useful physical quantity unless the distance can be determined or the magnitude compared with that of a star with known luminosity. Parallax, which is the apparent angular displacement of a star caused by the motion of the earth around the sun, is the most direct method of determining stellar distance (see Figure 2–1). Unfortunately, parallax measurements are possible only for objects whose distances are less than about 100 pc. Less direct methods must be used for objects at greater distances.

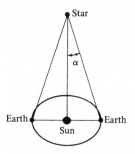

α = parallactic angle of star taken in plane of earth–sun orbit

Figure 2–1   The measurement of the parallactic angle of a star is illustrated.

The absolute visual magnitude M_v is defined to be the apparent visual magnitude m_v that an object as viewed through a standard yellow filter would have at a distance of 10 pc (1 pc = 3.26 light yr). Therefore, we have

$$M_v = m_v + 5 - 5 \log r + A \qquad (2\text{–}6)$$

where the distance r is in parsecs, and A is a correction factor for interstellar absorption. The absolute bolometric magnitude is defined as

$$M_b = M_v + B.C.$$

$$= -2.5 \log \left(\frac{L}{L\odot} \right) + 4.72 \qquad (2\text{–}7)$$

where

$$B.C. = -2.5 \log \frac{(\text{total luminosity})}{(\text{measured luminosity})},$$

$L\odot = 3.9 \times 10^{33}$ ergs/sec and 4.72 is the absolute bolometric magnitude of the sun.

A primary goal of the theory of stellar evolution is to predict how a star of given mass and initial chemical composition will evolve as a function of time and then compare these predictions with observations. The luminosity and effective temperature are two basic theoretically predicted properties of a star. During normal evolutionary phases most of the energy emitted by a star is in the form of photons. The effective temperature is the temperature of a blackbody having the same energy radiated per unit area as the star and therefore is defined by the expression

$$L = 4\pi R^2 \sigma T_e^{\ 4} \tag{2-8}$$

where $\sigma = ac/4 = 5.67 \times 10^{-5}$ ergs-sec^{-1}-cm^{-2}-deg^{-4}. Before a comparison between theory and observation can be made, the theoretically predicted quantities L and T_e (or R) must be related to the measured distance and observed apparent magnitude and temperature.

Stellar model calculations predict the effective temperature of a star. However, the effective temperature is not measured directly, and consequently it must be related to a measured surface temperature. Because the emitted spectrum of a star is not that of a perfect blackbody, the transformation from measured surface temperature to effective temperature is not trivial. At least three temperature scales are used to define surface temperature. These include color temperature, ionization temperature, and excitation temperature. The color temperature is defined by comparing the response as measured by means of a standard yellow (that is, visual) filter and blue filter with that predicted for a blackbody spectrum.

The excitation temperature is defined for states of the same ionization by the Boltzmann relation

$$\frac{n_k}{n_i} = \frac{g_k}{g_i} \exp\left[\frac{(E_i - E_k)}{kT}\right] \tag{2-9}$$

where g_k and g_i are the degeneracies of the k and i levels, respectively. Measurements of line strengths combined with known atomic transition probabilities allow the excitation temperature to be determined.

The ionization temperature is defined by comparing the observed strengths of absorption lines that arise from two different states of ionization of the same atom (for example, the H and K lines of neutral and singly ionized calcium) with that expected for a temperature T. If local thermodynamic equilibrium is assumed to exist, the Saha equation (see Appendix C) predicts

$$\frac{n_{r+1} n_e}{n_r} = \frac{g_{r+1} g_e}{g_r} \left(\frac{2\pi m k T}{h^2}\right)^{3/2} \exp\left(\frac{-X_r}{kT}\right) \tag{2-10}$$

where n_{r+1}/n_r is the ratio of abundances in two successive states of ionization; n_e is the number density of electrons; X_r is the ionization energy of the rth ionization state; and the g's are the degeneracy factors. The spectral classification of a star (that is, whether it is called type O, B, A, F, G, K, M) is an empirical temperature scale based on the presence and relative strengths of various spectral lines. The surface temperature of a star decreases monotonically as we go from type O to type M.

2–2 HERTZSPRUNG-RUSSELL DIAGRAM

The Hertzsprung-Russell (H-R) diagram represents stars according to absolute visual magnitude versus spectral type or color temperature. Luminous stars are near the top of the diagram, and relatively faint ones near the bottom. Hot stars are located toward the left, and relatively cool stars toward the right. The most striking feature of the H-R diagram shown in Figure 2–2 is that stars are found to preferentially occupy certain relatively narrow regions. This remarkable circumstance, which was recognized in the early part of this century, forms the basis for the development of the theory of stellar evolution. The position of a star in the H-R diagram is usually the most important observational clue about its evolutionary state. For this reason, a large part of the theory of stellar evolution is concerned with predicting on the basis of the equations of stellar interiors how a star of given initial mass and chemical composition will evolve in the H-R diagram. Although stars are very complicated and inaccessible, the results of stellar evolutionary calculations are believed to be basically correct, because with relatively simple physical assumptions, we can construct stellar models whose predicted evolution agrees (at least in broad outline) with the evolution of stars as inferred from observational evidence.

The three major classes of stars on the H-R diagram are main-sequence stars, red giants, and degenerate objects (white dwarfs, neutron stars, and black holes). Main-sequence stars are approximately chemically homogeneous and are responsible for most of the emitted visual radiation. They occupy

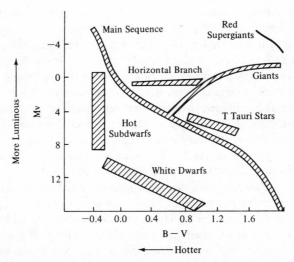

Figure 2–2 Some major classes of stars are shown in the H-R diagram. B-V is the standard blue magnitude minus the standard visual magnitude.

the diagonal strip on the H-R diagram that extends from the lower right-hand corner toward the upper left-hand corner. Massive main-sequence stars are relatively luminous and blue as compared to main-sequence stars of lower mass. If we compare a number of galactic clusters shown schematically in Figure 2–3, we find that although the various main sequences are approximately parallel on the H-R diagram, the limiting luminosities are higher for some clusters than for others. As will become clear below, this observation implies that the ages and maximum masses of main-sequence stars vary from cluster to cluster.

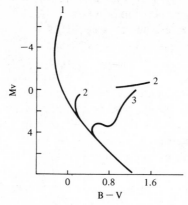

Figure 2–3   Curves 1, 2, and 3 denote galactic clusters of increasing age in the H-R diagram.

Stellar evolutionary calculations show that a star remains in the vicinity of the main sequence until a certain critical fraction of its hydrogen is transformed into helium (about 12 percent). After this limit, which is called the Schönberg-Chandrasekhar limit (see Section 5-3), is reached, the radius of the star increases by a large factor, and it becomes a red giant. Although the envelopes of red giants are very tenuous, their cores are dense.

White dwarfs are degenerate stars that represent the final evolutionary states of most stars. They occupy the lower left-hand corner of the H-R diagram. Neutron stars are very dense equilibrium configurations that are likely to be formed as a result of supernova explosions. Their interior densities probably exceed those inside normal nuclei ($\simeq 3 \times 10^{14}$ g-cm^{-3}). The existence of neutron stars has been inferred by the discovery of nonthermal sources called pulsars. Black holes are stellar configurations undergoing perpetual gravitational collapse. Although there is still no direct observational evidence for their existence, the formation of black holes is indicated because the masses of dense equilibrium configurations cannot exceed 1–2 M$\odot$.

Although stars are usually regarded as isolated systems, a sizable fraction (~ 25 percent) are members of binary systems. The presence of binary systems is of considerable importance. With the exception of the sun's mass, all direct knowledge of stellar masses is based on the study of binary systems. Moreover, the components of many binary systems are sufficiently close that mass exchange is likely to play a major role in the evolution of the system.

2–3   STELLAR MASSES

Since the history of a star is determined primarily by its initial mass and chemical composition, the accurate determination of these quantities is of great importance. The mass of the sun ($M\odot = 1.985 \times 10^{33}$ g) is, of course, the most precisely determined stellar mass. As mentioned above, direct determination of stellar masses is possible only for certain binary systems. Less certain theoretical arguments must be used to estimate the mass of most stars.

The methods of positional astronomy can be used to find the masses of the components of some binary systems. A determination of the orbit allows us to measure the sum of the masses by means of Kepler's third law, that is,

$$\frac{a^3}{P^2} = \frac{G}{4\pi^2} (M_1 + M_2) \tag{2–11}$$

where P is the period of revolution, and a is the semimajor axis of the orbit. The parallax of the orbit must be determined in order that a be found. In addition, if we measure the displacement of each of the two components with respect to a group of field stars, we can determine the ratio of the masses,

$$\frac{M_1}{M_2} = \frac{a_2}{a_1} \tag{2–12}$$

where a_1 and a_2 are the semimajor axes of the two measured ellipses relative to the center of mass. Equations (2–11) and (2–12) enable us to find the masses of the binary components since $a = a_1 + a_2$ (see Figure 2-4).

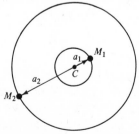

Figure 2-4   A binary system with circular orbits. M_1 and M_2 denote the masses of the components.

In certain instances, the masses of binary components too distant to be resolved visually can be determined by spectroscopic means. As we shall show below, the masses of spectroscopic binaries can be found only if the velocity curves of each component can be observed and the binary is an eclipsing variable such that the inclination of its orbit can be estimated.

For simplicity we assume that the orbits are circular. Since the time for each star to complete one revolution about the center of mass of the binary system is the same, we have

$$v_1 = \frac{2\pi a_1}{P}, \qquad v_2 = \frac{2\pi a_2}{P} \tag{2-13}$$

where v_1 and v_2 are the velocities of the components; a_1 and a_2 are the radii of the orbits about the center of mass; and P is the period of revolution. The angle of inclination i is defined to be the angle between the normal to the plane of the orbits, and the direction defined by a line drawn from the center of mass to an observer. If P, $v_1 \sin i$, and $v_2 \sin i$ can be measured, it follows from Equations (2–12) and (2–13) that

$$\frac{v_1 \sin i}{v_2 \sin i} = \frac{a_1}{a_2} = \frac{M_2}{M_1} \tag{2-14}$$

and consequently the ratio of the masses of the two components can be determined if their radial velocities are measured. From Equations (2–11), (2–13), and (2–14), we find

$$(M_1 + M_2) \sin^3 i = \frac{4\pi^2 P}{G} (v_1 \sin i + v_2 \sin i)^3. \tag{2-15}$$

Since the quantities on the right-hand side of Equation (2–15) can be measured, Equations (2–14) and (2–15) can be used to determine M_1 and M_2 if the binary is an eclipsing binary for which $i \simeq 90°$.

Measurements of stellar masses and luminosities determine an empirical mass-luminosity relation for main-sequence stars. This relation can be conveniently expressed in the form

$$L = CM^x \tag{2-16}$$

where C is a constant, and $3 \lesssim x \lesssim 4.5$ for stellar masses in the range $1 \, M\odot \le M \le 10 \, M\odot$. The theoretical basis for Equation (2–16) is discussed in Section 2–8. It follows from Equation (2–16) that the lifetimes of massive main-sequence stars are less than those with lower mass. Since massive main-sequence stars are relatively blue, young galactic clusters will contain bluer, more massive main-sequence stars than contained in older clusters. This latter conclusion depends on the plausible assumption that stars in a single cluster are formed at about the same time.

If we wish to explain the evolution of stars in the solar neighborhood or in a particular cluster, we must know how many stars are formed as a function of mass. The birth rate function $\psi(M)$ can be defined by the expression

$$dN = \psi(M)\, dM \qquad (2\text{--}17)$$

where dN is the number of stars per unit volume formed with masses between M and $M + dM$. $\psi(M)$ has been empirically estimated to be

$$\psi(M) = KM^{-2.3} \qquad (2\text{--}18)$$

where K is a constant for stars in the neighborhood of the sun. Equation (2–18), which has a low-mass cutoff at about 0.15 M$\odot$, implies that most stellar mass is in the form of low-mass stars. The luminosity function, which is defined as the number of stars per unit volume per unit luminosity interval, can be estimated if the birth rate function and mass-luminosity function are known.

2–4   STELLAR ATMOSPHERES AND CHEMICAL COMPOSITION

Important information about the history of matter and stellar evolution can be obtained by means of the spectroscopic determination of stellar abundances. The physical properties of a stellar atmosphere are determined by its surface gravity g, effective temperature T_e, chemical composition, and turbulent velocity. The theory of stellar atmospheres allows one to infer these quantities from the measured energy distribution of the spectrum. It follows from Equation (2–8) and the definition of the surface gravity,

$$g = \frac{GM}{R^2}, \qquad (2\text{--}19)$$

that the M/L ratio, but not the value of M or L, can be found from the theory of stellar atmospheres. The distance to a star must be known if M or L are to be independently determined.

In constructing model stellar atmospheres, one assumes hydrostatic equilibrium, a constant radiative luminosity, and plane parallel symmetry. The latter assumption is justified because the depth of the atmosphere is $\ll$ the radius of the star. It is convenient to express the equation of hydrostatic equilibrium in the form

$$\frac{dP}{d\tau} = \frac{g_e}{\kappa}, \qquad g_e = g - g_r \qquad (2\text{--}20)$$

where the opacity $\kappa = \kappa(T, P)$, $d\tau = \kappa\, dz$, and for the case of a gray atmosphere, $g_r = \kappa\sigma T_e^4/c$. The gravity g can be assumed constant throughout the atmosphere. The state of ionization of the gas is determined by means of the Saha equation (see Appendix C). The expression for the equation of state, that is,

$$\rho = \rho \, (T, \, P, \, \text{composition}) \tag{2-21}$$

allows one to determine the density as a function of the temperature, pressure, and chemical composition.

In order to integrate Equation (2-20), it is necessary to know how T varies as a function of τ. If the opacity is approximately independent of wavelength (for example, H^- opacity in the outer layers of the sun), then the Eddington approximation discussed in Appendix B is reasonably good. A knowledge of how T depends on τ allows one to calculate the effective temperature. More sophisticated mathematical methods are required if the opacity is strongly wavelength dependent. Mihalas[1] gives an excellent and up-to-date discussion of the theory of stellar atmospheres.

Values for g and T_e are usually poorly known, and it is only possible to measure directly how the observed intensity of radiation depends on wavelength. Therefore, in determining a model stellar atmosphere, the effective temperature, surface gravity, chemical composition, and turbulent velocity, which determine how much opacity is caused by lines, are varied until a satisfactory agreement between the observed and predicted dependence of the flux on wavelength has been obtained.

Although physical conditions on stellar surfaces or mixing of synthesized matter from stellar interiors are likely to be important in some stars, most measured stellar abundances are believed to be representative of interstellar abundances at the time of a star's formation. This circumstance implies that the determination of chemical abundances in stars of known evolution can help determine the chemical history of the universe.

The helium abundance in old stars is of special interest because it may be representative of abundances prevalent during the early history of the universe. The primordial fireball interpretation of the observed microwave background radiation suggests that ~ 30 percent by mass of the matter in the early universe was synthesized into helium. There is evidence to support this predicted helium production from theoretical investigations of RR Lyrae variables and horizontal branch stars. Moreover, the only known planetary nebula in a globular cluster (M15) has a helium abundance of about 30 percent by mass. On the other hand, spectroscopic investigations of old, hot stars indicate that a sizable fraction of these stars have low-helium abundances. These measurements suggest a low-helium abundance for old stars. However, it is possible that these measurements are indicative of abundances in the

---

[1] Mihalas, *Stellar Atmospheres*. San Francisco: Freeman, 1970.

outer layers only. The discovery that some hot, young stars also appear to be helium deficient strengthens the point of view that the observed low-helium abundances are representative of the surface layers only.

Because it is highly unlikely that elements heavier than Li^7 can be synthesized by means of the primordial fireball, it has been generally assumed that they are synthesized in the interiors of stars. This explanation for heavy-element formation is supported by the well-known result that in many old stars, the relative abundances of heavy elements such as carbon and barium are more than 100 times less than present estimates for the interstellar medium. It is of great interest to determine whether or not population I shows evidence for increasing heavy-element abundance and, in addition, to decide if the abundance ratio of one element to another changes with time. There is good evidence that the abundances of heavy elements increased markedly during the early history of the galaxy. However, evidence for a continuing, substantial increase is less conclusive.

Although heavy-element abundances are generally much lower for very old stars than for younger stars, the observed ratio of one heavy element to another is about the same for young and old stars. Moreover, these abundance ratios appear to be similar in external galaxies such as M31. This remarkable result has suggested a universal origin (that is, similar physical conditions) for the production of most heavy elements that eventually contaminate interstellar space.

It is not certain that most observed heavy elements were formed under conditions such as now exist in our galaxy. However, there is good evidence that nucleosynthesis is currently taking place in stellar interiors. Large numbers of hydrogen-deficient stars are observed. Moreover, the heavy-element technetium whose half-life is $\sim 2 \times 10^5$ yr has been detected in a number of red giant envelopes.

2–5 EQUATION OF STATE

Stellar matter can be approximated as a perfect gas because the average Coulomb energy per particle is much less than the average kinetic energy per particle, that is,

$$\frac{Ze^2}{\bar{r}} \ll \frac{3}{2} kT \tag{2–22}$$

where $\bar{r}$ is the mean interparticle distance, and Z is the charge of the positive ions. Condition (2–22) is satisfied in stars but not in ordinary matter because stellar matter is ionized except in the outermost layers where the density is very low.

If the gas is assumed to consist of a number of different fully ionized nuclei of atomic number A_i and mass fraction X_i, the number density of nuclei is

$$\sum_i n_i = \sum_i \frac{X_i}{A_i} \rho N_0 \qquad (2\text{-}23)$$

where $N_0 = 6 \times 10^{23}$ is Avogadro's number; ρ is the mass density; and the summation is over all nuclear species. The corresponding electron number density is

$$n_e = N_0 \rho \sum_i \frac{X_i Z_i}{A_i} \qquad (2\text{-}24)$$

where Z_i is the electron charge of nuclear species i. The mean molecular weight μ_m is defined by the expression

$$\mu_m^{-1} = \frac{n}{\rho N_0} = \sum_i \frac{X_i}{A_i}(1 + Z_i) \qquad (2\text{-}25)$$

where n is the total particle number density. Likewise, the mean electron molecular weight is

$$\mu_e^{-1} = \frac{n_e}{\rho N_0} = \sum_i \frac{X_i Z_i}{A_i}. \qquad (2\text{-}26)$$

It follows from Equation (2–25) and (2–26) that $\mu_m = \frac{1}{2}$ and $\mu_e = 1$ for ionized hydrogen, and $\mu_m = \frac{3}{4}$ and $\mu_e = 2$ for fully ionized helium.

The gas pressure of a perfect, nondegenerate gas is

$$P_g = \left(n_e + \sum_i n_i \right) kT = nkT. \qquad (2\text{-}27)$$

The total pressure P is the sum of the gas pressure and the radiation pressure, which is

$$P_R = \tfrac{1}{3}aT^4. \qquad (2\text{-}28)$$

If hydrogen molecules are completely dissociated and local thermodynamic equilibrium is established, the hydrogen and helium ionization equilibrium is described by means of the reactions:

$$\begin{aligned} \mathrm{H} &\rightleftarrows \mathrm{H}^+ + e^- \\ \mathrm{He} &\rightleftarrows \mathrm{He}^+ + e^- \\ \mathrm{He}^+ &\rightleftarrows \mathrm{He}^{++} + e^-. \end{aligned} \qquad (2\text{-}29)$$

The number densities of atoms, ions, and electrons are related by means of the Saha equation [see Equation (2–10)]. Some basic properties of the stellar equation of state are summarized in Figure 2–5.

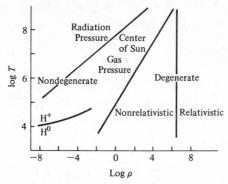

Figure 2–5   Some physical characteristics of the equation of state of a star are shown as a function of log T and log ρ.

Electron Degeneracy

Because electrons have half-integral spin, they obey Fermi-Dirac statistics and are called fermions. For fermions, each quantum state must be either occupied by a single particle or unoccupied. The probability f that an electron energy state is occupied is

$$f = \frac{1}{[\exp((E - \mu)/kT) + 1]} \qquad (2\text{–}30)$$

where μ is the chemical potential. If the ratio μ/kT is greater than unity, the gas is said to be degenerate, and $\mu \equiv E_F$ is called the Fermi energy. For the case of strong degeneracy ($E_F \gg kT$), all available electron states are occupied up to energy E_F since the function f is close to unity for $E < E_F$ and close to zero for $E > E_F$. The number density of electrons with momenta between p and $p + dp$ is obtained by multiplying f, which is the probability that a state will be occupied, by the total number density of states. We find

$$n_e(p)\, dp = 2f \frac{4\pi p^2}{h^3}\, dp \qquad (2\text{–}31)$$

where the factor of 2 arises because there exist two possible electron spin states for every momentum state.

We wish to derive an expression for the pressure of a perfect, degenerate gas. A particle of momentum p that is moving at an angle θ with respect to the normal to a surface element dA will transfer momentum

$$\Delta p = 2p \cos \theta \qquad (2\text{–}32)$$

upon reflection. The pressure on the surface is obtained by calculating the total rate of momentum transfer per unit area to the surface element, due to all the particles in the gas. Since the momentum distribution is isotropic, the number density of particles with momentum p that approach the surface element within the solid angle $2\pi \sin \theta \, d\theta$ is

$$n(p) \, dp \, \frac{2\pi \sin \theta \, d\theta}{4\pi}. \tag{2-33}$$

We construct a parallelepiped whose long axis has length v and is oriented in the θ direction, and whose lower and upper surfaces have area dA (assumed square). Since all particles inside the parallelepiped and moving with speed v in the θ directions will collide with the surface in a unit time, the pressure on the surface element dA becomes

$$P = \int_0^{\pi/2} \int_0^{\infty} (2p \cos \theta)(v \cos \theta) n(p) \, dp \, \frac{\sin \theta \, d\theta}{2}. \tag{2-34}$$

It follows from Equations (2–30), (2–31), and (2–34) that the pressure of a perfect electron gas is

$$P_e = \frac{8\pi}{3h^3} \int_0^{\infty} \frac{p^3 v \, dp}{[\exp((E - \mu)/kT) + 1]}. \tag{2-35}$$

2–6   EQUATIONS OF STELLAR INTERIORS

Hydrostatic and Thermal Equilibrium

Normal stars are generally in both hydrostatic and thermal equilibrium. Hydrostatic equilibrium arises when there is a balance between the attractive force of gravity and the outward force caused by the pressure gradient between inner and outer mass shells. If spherical symmetry is assumed, then only spherical mass shells need be considered. Consider a thin spherical shell of thickness dr and distance r from the center of the star. The downward gravitational force caused by gravity is

$$-\rho 4\pi r^2 \, dr \, \frac{GM_r}{r^2}$$

where

$$M_r = \int_0^r \rho 4\pi r^2 \, dr \tag{2-36}$$

is the mass interior to r. The net upward force on the spherical shell caused by pressure is

$$-\frac{dP}{dr} 4\pi r^2 \, dr.$$

Since the opposing forces caused by gravity and pressure must balance in hydrostatic equilibrium, we find

$$\frac{dP}{dr} = -\rho \frac{GM_r}{r^2}. \tag{2-37}$$

Thermal equilibrium is a consequence of an equilibrium between the rate of production and loss of energy. If contractional energy release can be neglected, the luminosity at a radius r is

$$L_r = \int_0^r 4\pi r^2 \varepsilon \rho \, dr \qquad (2\text{--}38)$$

where ε is the rate of nuclear energy generation in units of ergs-g^{-1}-sec^{-1}. A more general equation for the conservation of energy in stellar interiors will be discussed below. In order to complete our description of the interior an equation of energy transport must be obtained.

Radiative Equilibrium

The transport of energy in the interior of a star can be caused by radiative transport, convection, or electron conduction. Since the mean free path of a photon is generally much less than the stellar radius except in the outermost surface layers, the diffusion approximation is an accurate description of radiative energy transport in the stellar interior. Therefore, under steady-state conditions, the radiative energy flux becomes

$$F \equiv \frac{L_r}{4\pi r^2} = -D \text{ grad } U_r \qquad (2\text{--}39)$$

where U_r is the radiation energy density, and D the diffusion constant. It can be shown from kinetic theory that the diffusion constant is

$$D = \tfrac{1}{3}\lambda c. \qquad (2\text{--}40)$$

λ is the photon mean free path,

$$\lambda = \frac{1}{n\sigma} = \frac{1}{\kappa\rho} \qquad (2\text{--}41)$$

where σ is the cross section, and n is the number density of absorbers (or scatterers). The opacity κ, which is in units of cm^2-g^{-1}, is defined by the relation $n\sigma = \kappa\rho$. Since the energy density of radiation is

$$U_r = aT^4 \qquad (2\text{--}42)$$

it follows from Equation (2–39) that

$$\frac{L_r}{4\pi r^2} = -\frac{4}{3}\frac{acT^3}{\kappa\rho} \text{ grad } T \qquad (2\text{--}43)$$

is the equation of radiative energy transport in stellar interiors.

Because the opacity is usually a function of frequency, the above discussion must be modified. For each frequency v, we have

$$F_v = -D_v \text{ grad } U_v \qquad (2\text{--}44)$$

where

$$U_v = \frac{8\pi h v^3}{c^3(e^{hv/kT} - 1)} \tag{2-45}$$

and

$$D_v = \frac{1}{3}\frac{c}{\rho\kappa_v}. \tag{2-46}$$

Since

$$\frac{L_r}{4\pi r^2} = \int_0^\infty F_v\,dv = -\frac{c}{3\rho}\int_0^\infty \frac{1}{\kappa_v}\,\mathrm{grad}\,U_v\,dv \tag{2-47}$$

and

$$\mathrm{grad}\,U_v = \frac{dU_v}{dT}\,\mathrm{grad}\,T \tag{2-48}$$

we find

$$\frac{L_r}{4\pi r^2} = -\frac{c\,\mathrm{grad}\,T}{3\rho}\int_0^\infty \frac{1}{\kappa_v}\frac{dU_v}{dT}\,dv. \tag{2-49}$$

For blackbody radiation we have

$$\int U_v\,dv = aT^4 \tag{2-50}$$

and

$$T\int \frac{dU_v}{dT}\,dv = 4aT^4. \tag{2-51}$$

It follows that if we define the opacity by means of the expression

$$\frac{1}{\kappa} = \frac{\int (1/\kappa_v)\,(dU_v/dT)\,dv}{\int (dU_v/dT)\,dv} \tag{2-52}$$

the form of Equation (2–49) becomes identical with that of Equation (2–43). The opacity as defined in Equation (2–52) is called the Rosseland mean opacity.

Opacity

In order to calculate the Rosseland mean opacity, it is necessary to determine the relevant scattering and absorption cross sections as a function of photon energy. Electron scattering, free-free absorption, bound-free absorption (also called photoionization), and bound-bound absorption are the principal sources of opacity in stars.

The cross section for the scattering of low-energy photons by means of electrons (called Thomson scattering) can be readily calculated from classical electromagnetic theory. The power radiated from an accelerated non-relativistic electron is known to be

$$P = \frac{2}{3} \frac{e^2}{c^3} \left(\frac{dv}{dt}\right)^2 \quad \text{(ergs/sec)} \qquad (2\text{-}53)$$

where dv/dt is the acceleration of the electron. An incident electromagnetic wave with electric field strength

$$E = E_0 \sin \omega t \qquad (2\text{-}54)$$

will accelerate a free electron by an amount

$$\frac{dv}{dt} = -\frac{eE_0 \sin \omega t}{m}. \qquad (2\text{-}55)$$

Consequently Equation (2-53) becomes

$$P = \frac{2}{3} \frac{e^4 E_0^2 \sin^2 \omega t}{c^3 m^2}. \qquad (2\text{-}56)$$

The power per unit area transported by an electromagnetic wave is given by the Poynting vector

$$\mathbf{S} = \frac{c}{4\pi} (\mathbf{E} \times \mathbf{B}) = \frac{c}{4\pi} E_0^2 \sin^2 \omega t \, \hat{k} \qquad (2\text{-}57)$$

where $\hat{k}$ is a unit vector in the direction of propagation of the wave. From Equations (2-56) and (2-57), the Thomson scattering cross section becomes

$$\sigma_T = \frac{P}{S} = \frac{8\pi r_0^2}{3} = 0.665 \times 10^{-24} \text{ cm}^2 \qquad (2\text{-}58)$$

where $r_0 = e^2/mc^2$ is the classical radius of the electron. If the mass fraction of hydrogen is denoted by X, and the ratio of charge Z to atomic mass A is assumed equal to $\frac{1}{2}$ for elements other than hydrogen, then the opacity caused by Thomson scattering becomes

$$\kappa = \frac{n_e \sigma_T}{\rho} = 0.2(1 + X). \qquad (2\text{-}59)$$

The expression for opacity given in Equation (2-59) is independent of temperature and density. At high temperatures ($T \gtrsim 10^8 °\text{K}$), the classical Thomson cross section σ_T is no longer accurate because when the energy of the photon $h\nu$ becomes comparable to mc^2, the scattering is no longer isotropic, but occurs preferentially in the forward direction. For this reason, the opacity is reduced at high temperatures. Electron degeneracy will further reduce the opacity since an electron cannot be scattered into a quantum state that is already occupied by another electron.

Free-free absorption arises when the absorption of a photon causes an electron in one continuum state to be excited into a continuum state of higher energy. It is the inverse of bremsstrahlung, which occurs when a free electron emits a photon as it is accelerated by the electric field of an ion. The principle of detailed balance asserts that under conditions of thermodynamic equilibrium, the rate of any particular transition must be the same as its inverse. For this reason, the opacity caused by free-free absorption can be related to the rate of bremsstrahlung emission calculated in Section 1–3.

From Equation (1–22) it follows that the energy radiated in units of ergs-g^{-1}-sec^{-1} by means of thermal bremsstrahlung is proportioned to

$$\varepsilon \propto \frac{Z^2 n_e n_i T^{1/2}}{\rho} \tag{2–60}$$

where ρ is the density of the plasma, Z the charge of an ion, and n_e and n_i the electron and ion number densities, respectively. The corresponding amount of absorbed energy is

$$\kappa \sigma T^4$$

where κ is the free-free absorption opacity, and σ is the Stefan-Boltzmann constant. Equating the energy absorbed to that radiated, we find

$$\kappa \propto Z^2 n_e n_i T^{-7/2} \tag{2–61}$$

which is called Kramers' opacity law. Equation (2–61) can be rewritten as

$$\kappa \propto \frac{Z^3 \rho}{A^2} T^{-7/2} \tag{2–62}$$

where A is the atomic weight of the ion.

Bound-free absorption arises when the absorption of a photon by an atom causes a bound electron to make a transition to a continuum state. An expression for the bound-free absorption cross section for hydrogenic atoms is given in Equation (6–1). The density and temperature dependence of bound-free opacity is similar to that of free-free opacity.

Bound-bound absorption takes place when the absorption of a photon is accompanied by the transition of an electron from one bound state to another. The great importance of heavy elements in determining stellar opacity arises primarily from the circumstance that the cross section for bound-bound absorption is much greater than that for bound-free or free-free absorption. Therefore, in regions of a star where H and He are fully ionized but heavy elements such as iron are not, the opacity can be dominated by heavy elements even though their abundances are low. The calculation of bound-bound absorption, which in general is very complicated, constitutes the largest uncertainty in determining stellar opacity.

Convective Energy Transport

If the temperature gradient in a stellar interior becomes sufficiently large, the mechanical equilibrium will become unstable, and convective motions will result. Let the mass density of a fluid initially at rest be taken as a function of pressure and entropy density, $\rho = \rho(P, s)$. Let $P_0(r)$, $s_0(r)$, and $\rho_0(r)$ represent the equilibrium distribution of the pressure, entropy, and mass density, respectively. The condition for stability against convective motions can be derived as follows. A small element of fluid initially at equilibrium position r is displaced a distance δr. The displaced element is assumed to maintain pressure equilibrium with its surroundings. Therefore, we have

$$\rho'(r + \delta r) = \rho'(P_0(r + \delta r), s').\tag{2-63}$$

The initial equilibrium will be stable against convection if upwardly displaced elements sink and downwardly displaced elements rise. In general, it is sufficient to consider adiabatic perturbations (that is, $s = $ constant) since upwardly displaced elements usually lose heat while rising, and downwardly displaced elements gain heat. Therefore, a sufficient condition for convective stability is

$$\rho'(P_0(r + \delta r), s_0(r)) - \rho_0(P_0(r + \delta r), s_0(r + \delta r)) \quad \begin{cases} >0 & \text{if } \delta r > 0 \\ <0 & \text{if } \delta r < 0. \end{cases}$$

$$\tag{2-64}$$

If Equation (2–64) is expanded to first order in δr, we have

$$-\left(\frac{\partial \rho}{\partial s}\right)\frac{ds_0}{dr} > 0.\tag{2-65}$$

Since the heat capacity at constant pressure is

$$c_P = T\left(\frac{\partial s}{\partial T}\right)_P\tag{2-66}$$

we have

$$\left(\frac{\partial \rho}{\partial s}\right)_P = \left(\frac{\partial \rho}{\partial T}\right)_P \bigg/ \left(\frac{\partial s}{\partial T}\right)_P = \frac{T}{c_P}\left(\frac{\partial \rho}{\partial T}\right)_P.\tag{2-67}$$

If the chemical composition of the fluid is constant, Equations (2–65) and (2–67) imply

$$\frac{ds_0(r)}{dr} > 0\tag{2-68}$$

since

$$\left(\frac{\partial \rho}{\partial T}\right)_P < 0.$$

Condition (2–68) can be rewritten as

$$\frac{ds}{dr} = \left(\frac{\partial s}{\partial T}\right)_P \frac{dT_0}{dr} + \left(\frac{\partial s}{\partial P}\right)_T \frac{dP_0}{dr}$$

$$= \frac{c_P}{T}\frac{dT_0}{dr} + \frac{\delta}{\rho T}\frac{dP_0}{dr} \tag{2–69}$$

where

$$\delta = -\left(\frac{\partial \ln \rho}{\partial \ln T}\right)_P.$$

Using the equation of hydrostatic equilibrium to eliminate dP_0/dr from Equation (2–69), we find

$$\frac{dT_0}{dr} > -\frac{g\delta}{c_P} = \left(\frac{dT_0}{dr}\right)_s \tag{2–70}$$

for convective stability. Condition (2–70) shows that if the temperature decreases more rapidly as a function of r than the adiabatic gradient, convection will result. Under conditions of steady convection, only properly averaged values of the physical variables satisfy the equation of hydrostatic equilibrium. Although there are important exceptions, convection is usually very efficient in stellar interiors, and consequently the actual temperature in a convective region is nearly equal to the adiabatic temperature gradient.

Conservation of Energy

The first law of thermodynamics can be written

$$\Delta U = \Delta Q - \Delta W \tag{2–71}$$

where ΔU is the change in internal energy, ΔQ the heat added to the system, and ΔW the work done by the system. The net heat added per gram to a spherical shell of a star in the time interval Δt is

$$\Delta Q = \left(-\frac{dL_r}{dM_r} + \varepsilon_n - \varepsilon_v\right)\Delta t \tag{2–72}$$

where ε_n is the rate of nuclear energy generation, and ε_v the rate of neutrino emission.

The work done per unit mass by a mass element is

$$\Delta W = P \Delta \left(\frac{1}{\rho}\right). \tag{2–73}$$

From Equations (2–71), (2–72), and (2–73), we find

$$\frac{dU}{dt} + P\frac{d(1/\rho)}{dt} = \varepsilon_n - \varepsilon_v - \frac{dL_r}{dM_r}. \tag{2–74}$$

For the special case of a nonrelativistic, fully ionized, perfect gas we have

$$U = \frac{3}{2}\frac{P}{\rho} \qquad (2\text{-}75)$$

which implies that Equation (2-74) can be written in the convenient form

$$\tfrac{3}{2}\rho^{2/3}\frac{d(P/\rho^{5/3})}{dt} = \varepsilon_n - \varepsilon_\nu - \frac{dL_r}{dM_r}. \qquad (2\text{-}76)$$

Equations (2-36), (2-37), (2-74), and (2-43) or (2-70) are the equations of stellar interiors. The solutions to these equations, which must be solved by numerical means, determine the evolution of a star if physical boundary conditions are imposed at the surface and center. A primary goal of the theory of stellar evolution is to solve these equations and thereby predict the evolution of stars as a function of initial mass and chemical composition. The great interest in solving these equations rests in large part on the circumstance that except for differing amounts of angular momentum, the initial main-sequence states of stars are likely to be well determined since nuclear time scales are generally much longer than corresponding thermal time scales. If the initial state of a star is assumed known, then by solving the equations of stellar interiors, we can trace the evolution of stars in the H-R diagram and compare our predictions with observations. Moreover, by studying the physical properties of stellar models, we can hope to achieve a better understanding of stellar phenomena such as variable stars, mass loss, planetary nebulae, novae, galactic x-ray sources, and supernovae. In addition, a correct theory of the evolution of stars should provide a satisfactory theory for the formation of elements heavier than helium since such elements are likely to be formed in stellar interiors.

2-7 NUCLEAR REACTIONS IN STARS

The number of reactions per unit time per unit volume between two nuclear species with number densities n_1 and n_2 is

$$\frac{n_1 n_2 \langle \sigma v \rangle}{1 + \delta_{12}} = \frac{\rho \varepsilon_n}{Q} \qquad (2\text{-}77)$$

where v is the relative velocity of the colliding nuclei; Q is the energy release per reaction; σ is the reaction cross section; and δ_{12} is unity if the nuclei are identical and zero otherwise. The reaction rate per pair of particles is defined by the expression

$$\langle \sigma v \rangle = \int_{v_{\min}} \sigma v f(v)\, d^3 v \qquad (2\text{-}78)$$

where $v_{\min}$ is the relative particle velocity necessary for reaction threshold, and $f(v)\, d^3 v$ is the probability that the relative velocity of two particles is in the volume of velocity space, d^3v. If the velocity distribution of the particles is Maxwellian, we find

$$\langle \sigma v \rangle = \left(\frac{8}{\pi m_r} \right)^{1/2} (kT)^{-3/2} \int_{E_{\min}}^{\infty} E\sigma(E) \exp\left(-\frac{E}{kT} \right) dE \qquad (2\text{--}79)$$

where

$$m_r = \frac{m_1 m_2}{m_1 + m_2}$$

is the reduced mass of the colliding particles and E is the center of mass energy.

In order that two particles undergo a nuclear interaction, it is necessary that they penetrate the Coulomb barrier between them. The Coulomb interaction is approximately

$$V \sim \frac{Z_1 Z_2 e^2}{R}. \qquad (2\text{--}80)$$

The nuclear charge distribution extends over a radius $\sim 1.4 \times A^{1/3} \times 10^{-13}$ cm. Typical particle energies are of order kT for the Maxwell distribution. Since temperatures are usually in the range $2 \times 10^7 - 10^{8}$°K, most particles will have energies that are much less than the Coulomb barrier potential, and consequently barrier penetration will be very improbable except for particles in the high-energy tail of the Maxwell distribution. For low-energy interactions, it can be shown that the probability of penetrating the Coulomb barrier is proportional to

$$\exp\left(-\frac{2\pi Z_1 Z_2 e^2}{\hbar v} \right) \qquad (2\text{--}81)$$

which is called the Gamow factor.

The low-energy interaction cross section is proportional to the square of the de Broglie wavelength, that is,

$$\lambda^2 = \frac{h^2}{p^2} \propto \frac{1}{E}. \qquad (2\text{--}82)$$

Therefore, from Equations (2–81) and (2–82) it becomes plausible to write the nonresonant low-energy nuclear cross section in the form

$$\sigma(E) = \frac{S(E)}{E} \exp\left(-\frac{2\pi Z_1 Z_2 e^2}{\hbar v} \right) \qquad (2\text{--}83)$$

where $S(E)$ is a slowly varying function of energy. Expanding $S(E)$, we find

$$S(E) \simeq S(0) + \left(\frac{dS}{dE} \right)_0 E. \qquad (2\text{--}84)$$

If the terms $S(0)$ and $(dS/dE)_0$ can be found experimentally, then the reaction rate can be determined as a function of temperature and density. If the energies of the interacting particles are close to a nuclear resonance, the expression for $\sigma(E)$ given above must be replaced by the Breit-Wigner single-level cross section.

Electron shielding caused by electrons surrounding nuclei in stars increases stellar reaction rates as compared with those measured in the laboratory. Although this screening correction is generally small, it can become very large under conditions of very high density such as may arise during the ignition of helium and carbon core burning.

Principal Nuclear Burning Stages

The conversion of hydrogen into helium represents the most important energy source for normal stars. The proton-proton reaction chains and the CNO cycle are the two ways in which the conversion of hydrogen into helium can take place in stars.

The proton-proton reaction chains are initiated by either of the weak interactions,

$$p + p \rightarrow D^2 + e^+ + \nu_e$$

or   (2–85)

$$p + e^- + p \rightarrow D^2 + \nu_e \quad \text{(one out of about 400 reactions)}.$$

The cross section for these interactions in the interiors of stars is very low ($\sim 10^{-47}$ cm^2). The remaining reactions are

$$D^2 + p \rightarrow He^3 + \gamma$$

$$He^3 + He^4 \rightarrow Be^7 + \gamma \qquad He^3 + He^3 \rightarrow He^4 + 2p$$

$$Be^7 + e^- \rightarrow Li^7 + \nu_e \qquad Be^7 + p \rightarrow B^8 + \gamma \qquad (2\text{–}86)$$

$$Li^7 + p \rightarrow He^4 + He^4 \qquad B^8 \rightarrow Be^8 + e^+ + \nu_e$$

$$Be^8 \rightarrow 2\,He^4.$$

The energy release by means of the proton-proton reaction chains is $\simeq 6.5$ MeV per proton. About 2 percent of this nuclear energy release is emitted in the form of neutrinos.

Neutrino-capture cross sections are very low, and consequently neutrinos that are emitted in the interior of the sun pass through without interaction. The detection of solar neutrinos would constitute a means of directly studying the center of the sun. Although most solar neutrinos come from the proton-proton reaction, these neutrinos have low energy and are difficult to detect since the cross section for neutrino capture is proportional to the square of the energy. In principle, solar neutrinos emitted by the decay of B^8 should be of sufficient energy ($\simeq 7$ MeV) and flux to be detected by means of the reactions:

$$\nu + Cl^{37} \rightarrow Ar^{37} + e^- - 0.814 \text{ MeV}$$
$$Ar^{37} + e^- \rightarrow Cl^{37}* + \nu \qquad (2\text{-}87)$$
$$Cl^{37}* \rightarrow Cl^{37} + e^-.$$

The ejected $\simeq 0.28$ keV electron is detected. Recent measurements of solar neutrinos by means of a tank of 10^5 gallons of C_2Cl_4 placed 1 mile below the ground indicate lower neutrino fluxes than had been expected, and consequently the inferred central temperature of the sun is about 5–10 percent less than had been predicted, or that one or more of the estimated nuclear reaction rates is in error. It can also be concluded that less than 5 percent of the sun's emitted radiation is generated as a result of the CNO cycle. This latter result is independent of stellar structure calculations. Some physical properties of the solar interior are shown schematically in Figure 2–6.

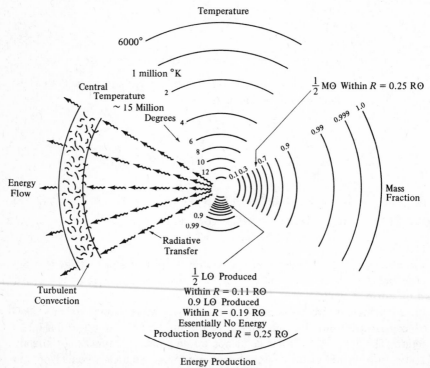

Figure 2-6   Some physical properties of the solar interior are shown schematically (based on calculations by Iben).

The CNO cycle becomes the dominant means of converting hydrogen into helium at temperatures in excess of $\sim 16 \times 10^{6\circ}$K for stars with solar abundances. The dominant mode of completing the CNO cycle is

$$C^{12} + p \rightarrow N^{13} + \gamma$$
$$N^{13} \rightarrow C^{13} + e^+ + \nu_e \quad \text{(mean lifetime = 870 sec)}$$
$$C^{13} + p \rightarrow N^{14} + \gamma \quad\quad\quad\quad\quad\quad (2\text{-}88)$$
$$N^{14} + p \rightarrow O^{15} + \gamma$$
$$O^{15} \rightarrow N^{15} + e^+ + \nu_e \quad \text{(mean lifetime = 178 sec)}$$
$$N^{15} + p \rightarrow C^{12} + He^4.$$

It can be shown that if the CNO cycle has time to reach equilibrium conditions most of the initial CNO elements are converted into N^{14}. Moreover, the predicted equilibrium C^{13}/C^{12} ratio is $\sim \frac{1}{4}$. Both the relatively high solar and interstellar abundances of C^{12} and O^{16} and the corresponding observed C^{13}/C^{12} ratio ($\sim \frac{1}{80}$) indicate that most of the C^{12} and O^{16} in stars are produced at a later evolutionary stage. However, there are stars whose surface abundances are indicative of CNO equilibrium abundances. Unlike the proton-proton reaction chains that are not very temperature sensitive

$$(\varepsilon_{pp} \sim T^n, n \sim 4),$$

the CNO cycle is quite temperature sensitive

$$(\varepsilon_{CNO} \sim T^n, n \sim 15),$$

and consequently main-sequence stars that operate on the CNO cycle have convective cores.

After hydrogen core exhaustion, a star contracts, and its central temperature increases. As the core contracts, a hydrogen burning shell source forms, and the star evolves up the red giant branch. The central temperature continues to increase as the star evolves up the red giant branch until helium burning commences at a temperature of about $10^{8\circ}$K. Helium burning takes place as a result of the triple α reactions:

$$2\,He^4 \rightleftarrows Be^8 + \gamma$$
$$Be^8 + He^4 \rightarrow C^{12} + \gamma \quad\quad\quad (2\text{-}89)$$
$$C^{12} + He^4 \rightarrow O^{16} + \gamma.$$

The triple α reactions are expected to produce comparable amounts of C^{12} and O^{16}. This circumstance explains why the observed abundances of C^{12} and O^{16} are nearly the same.

Carbon burning is expected to follow helium burning. The total mass of two C^{12} nuclei is ~ 14 MeV above the ground state of Mg^{24}, and consequently a large number of compound nuclear states will contribute to the reaction rate. Carbon burning will be initiated at $T \sim 4\text{--}7 \times 10^{8}\,°K$. The following decay modes of Mg^{24} are expected:

$$
\begin{array}{lr}
 & Q\ (MeV) \\
C^{12} + C^{12} \to Mg^{24*} \to Ne^{20} + \alpha & 4.0 \\
\qquad\qquad\qquad \to Na^{23} + p & 2.2 \\
\qquad\qquad\qquad \to Mg^{24} + \gamma & 13.9 \\
\qquad\qquad\qquad \to Mg^{23} + n & -2.6 \\
\qquad\qquad\qquad \to O^{16} + 2\alpha & -0.114
\end{array}
\tag{2-90}
$$

The first two decay modes are the most important ones in the interiors of stars, and consequently Ne^{20} and Na^{23} are the primary direct products of carbon burning. However, the final reaction products of carbon burning can be found only if reactions of the α particles and protons are included with each of the direct reaction products.

Carbon burning can lead to the production of an intense neutron flux by means of the reactions

$$
\begin{aligned}
p + C^{12} &\to N^{13} \\
N^{13} &\to C^{13} + e^{+} + \nu_e \\
\alpha + C^{13} &\to O^{16} + n.
\end{aligned}
\tag{2-91}
$$

The interaction of these neutrons with seed nuclei such as Fe^{56} can lead to the formation of heavy elements.

Oxygen burning is expected to follow carbon burning at temperatures $\sim 1.5 - 2 \times 10^{9}\,°K$. S^{28} is the principal end product of oxygen burning because it is the most tightly bound isotope produced. One might expect that reactions such as Mg burning would follow oxygen burning. However, photodisintegration of nuclei will take place at lower temperatures than are necessary for these reactions.

2–8   DIMENSIONAL ANALYSIS AND STELLAR MODELS

Relative Magnitudes of Physical Variables

Many of the most important properties of chemically homogeneous stellar models can be demonstrated by means of dimensional analysis. It can be shown that if the nuclear energy source is specified and the equation of state is that of an ideal, nondegenerate gas, the structure of a star is determined by the mass M, the molecular weight μ, and the conditions of hydrostatic and thermal equilibrium.

If we replace dP/dr by P/R in Equation (2–37) and substitute for ρ the relation

$$
\rho \sim \frac{M}{R^{3}}
\tag{2-92}
$$

we find

$$P \sim \frac{GM^2}{R^4}. \tag{2-93}$$

Substituting the equation of state for a nondegenerate gas (that is, $P = \rho kT/\mu m_H$) into Equation (2-93) gives the additional relation

$$T \sim \frac{Gm_H\mu M}{kR} \tag{2-94}$$

where we have neglected radiation pressure.

If radiative energy transport is assumed, we can write

$$L \sim \frac{ac}{3} 4\pi R^2 \frac{T^4}{R} \frac{1}{\kappa\rho} \sim \frac{ac}{3} 4\pi \left(\frac{m_H G}{k}\right)^4 \frac{\mu^4 M^3}{\kappa} \tag{2-95}$$

where we have replaced dT/dr by T/R in Equation (2-43) and have used Equation (2-94) to obtain the second expression in Equation (2-95). It is physically more illuminating to obtain Equation (2-95) by means of the following argument. The radiative luminosity is approximately

$$L \sim \frac{\text{total photon energy}}{\text{photon diffusion time}} = \frac{(4\pi/3)R^3 aT^4}{\tau_K}. \tag{2-96}$$

The random walk (diffusion) approximation implies that $\tau_K = (R/c)(R/\lambda)$, where c is the speed of light, λ the mean free path of a photon, and R the characteristic dimension of the region. Since the photon mean free path is $\lambda = 1/(\kappa\rho)$, Equation (2-96) is seen to be equivalent to Equation (2-95).

If the opacity is primarily caused by electron scattering, $\kappa = 0.2(1 + X)$, and Equation (2-95) implies that $L \propto G^4\mu^4 M^3$. On the other hand, if bound-free (Kramers') opacity is dominant, $\kappa = \kappa_0 \rho T^{-3.5}$ and Equation (2-95) gives $L \propto G^{7.5}\mu^{7.5}M^{5.5}/R^{0.5}$. Electron scattering is the principal source of opacity for massive main-sequence stars since their densities are relatively low. The above mass-luminosity relations show that the luminosity of a homogeneous star is strongly dependent on its mass and molecular weight as well as the gravitational constant.

The strong dependence of the luminosity on the gravitational constant is of considerable importance to cosmology since certain cosmological theories require that the value of G be decreasing as the universe evolves. Geological evidence indicates that the age of the sun is greater than 4.5×10^9 yr. It can be shown that if the true value of G changed by as much as a factor of 2 during the last 4.5×10^9 yr, the sun would have already exhausted its central hydrogen supply and evolved off the main sequence.

If gravitational energy release can be neglected, the condition that a star be in thermal equilibrium implies that its luminosity equal the total nuclear energy generation,

$$L \sim \varepsilon M \tag{2-97}$$

where ε is the rate of nuclear energy generation in ergs-g^{-1}-sec^{-1}. It is convenient to write ε in the form $\varepsilon = \varepsilon_0 \rho^\beta T^\nu$, where β and ν indicate the density and temperature dependences, respectively, of the nuclear reactions. The latter expression for ε and Equation (2–97) give

$$L \sim \varepsilon_0 \left(\frac{Gm_H}{k}\right)^\nu \frac{\mu^\nu M^{\nu+\beta+1}}{R^{\nu+3\beta}}. \tag{2–98}$$

Typical values for (β, ν) are (1, 4) for the proton chain, (1, 15) for the CNO cycle, and (2, 30) for the triple α reactions. Equations (2–95) and (2–98) provide a relation between the mass and radius of a star if the molecular weight μ is specified. Equations (2–92), (2–93), (2–94), (2–95), and (2–98) determine the basic physical variables of a star (that is, P, ρ, T, R, and L) as a function of its mass and molecular weight. A more mathematically complete derivation of the statement that the structure of a star is determined by its mass and molecular weight is known as the Vogt-Russell theorem.

Stellar Stability

Dimensional analysis can be used to study the dynamical and thermal stability of a simple stellar model. Equations (2–92) and (2–93) give the relation

$$P \sim GM^{2/3}\rho^{4/3}. \tag{2–99}$$

We wish to examine the stability of the equilibrium implied by Equation (2–99). Since the thermal diffusion time (that is, Kelvin time) is in general much longer than the dynamical time scale, the dynamical stability of the configuration must be studied at constant entropy S. For the configuration to be dynamically stable, the pressure must increase faster under compression than the equivalent gravitational pressure. Equation (2–99) shows that the star will be dynamically stable if

$$\gamma \equiv \frac{d \ln P}{d \ln \rho} > \frac{4}{3}. \tag{2–100}$$

If we allow M to vary and differentiate Equation (2–99) with respect to ρ, we find

$$\frac{dM}{d\rho} \sim \frac{P\rho^{-7/3}M^{1/3}}{G}\left[\frac{d \ln P}{d \ln \rho} - \frac{4}{3}\right]. \tag{2–101}$$

Equation (2–101) shows that the onset of dynamical instability will occur when $dM/d\rho = 0$. This result will be of considerable importance when we discuss neutron stars.

Both the CNO cycle and triple α reactions depend much more sensitively on temperature than the radiative energy losses which vary as T^4. For this reason, it would appear plausible that stars which are supported by these nuclear reactions should be thermally unstable. However, most such stars are not thermally unstable because $\tau_D \ll \tau_k$, and they consequently can expand to compensate for excess nuclear energy release. Equation (2–99) shows that for a chemically homogeneous star, the requirement of hydrostatic equilibrium implies that the expansion that must accompany an excess release of nuclear energy will act to decrease the temperature and thereby stabilize the star. Since the addition of thermal energy decreases the temperature, and conversely, we can say that the star behaves as if its heat capacity were negative. A star is an open system whose emitted radiation transports entropy to infinity, and therefore the above behavior does not contradict the second law of thermodynamics. Not all stars are expected to be thermally stable as we shall see.

Upper and Lower Mass Limits

It is a conspicuous fact that most of the observable mass in our galaxy is contained in stars whose masses lie in the range ~ 0.2 M$\odot$–100 M$\odot$. In this section we shall discuss some plausible reasons for this observation.

In a crude way the upper limit for stable main-sequence stars is very likely related to the dominance of radiation pressure as compared to gas pressure for massive stars. The average thermal gas energy per particle is

$$u_{\text{gas}} = \tfrac{3}{2}kT \tag{2–102}$$

and the radiation energy per particle for ionized hydrogen is

$$u_{\text{photon}} \sim aT^4 \left(\frac{\tfrac{4}{3}\pi R^3}{2N} \right) \tag{2–103}$$

where $N = M/m_p$ is the total number of nucleons. The virial theorem tells us that

$$u_{\text{gas}} + u_{\text{photon}} \sim \frac{Gm_p^2 N}{R} . \tag{2–104}$$

Equations (2–102), (2–103), and (2–104) imply

$$\frac{u_{\text{photon}}}{(u_{\text{gas}})^4} \propto \frac{N^2}{(u_{\text{gas}} + u_{\text{photon}})^3} . \tag{2–105}$$

The above relation shows that for sufficiently massive stars, the internal energy is primarily due to radiation. The virial theorem implies that a star dominated by a relativistic gas such as photons will be nearly gravitationally unbound. Since the gas pressure is small as compared to the radiation pressure, we can write

$$P \sim \frac{a}{3} T^4 \sim \left(\frac{3S^4}{256a} \right)^{1/3} \rho^{4/3} \tag{2–106}$$

where the entropy $S \sim 4aT^3/3\rho$ must be nearly constant throughout because most of the star will be unstable to convection and the temperature gradient will be nearly adiabatic. Since electron scattering is the principal source of opacity, it follows from Equations (2–92), (2–93), (2–96), and (2–106) that

$$\frac{L}{L\odot} \sim \frac{10^5 M}{1.5(1 + X)\, M\odot} \tag{2-107}$$

for very massive stars.

The above argument does not really imply an upper-mass limit since there is some net binding even for arbitrarily large mass if the effect of general relativity is neglected. Observations show that the very luminous star η Carinae is ejecting shells of mass. Its luminosity, which is concentrated in the infrared, is $\simeq 10^6$ L$\odot$, and therefore its mass may exceed 100 M$\odot$. Other observational evidence concerning the upper-mass limit comes from binary system observations. These observations do not indicate the existence of stars more massive than about 50 M$\odot$ in binary systems. Although it can be shown that main-sequence stars more massive than about 100 M$\odot$ should be pulsationally unstable [see Equation (4–11)], hydrodynamic calculations show that nonlinear effects are likely to limit the amplitude of pulsations before they can reach amplitudes sufficient to disrupt the entire star. It appears likely that other effects such as may arise during the formation of massive stars may lead to an upper-mass limit (see Chapter 3).

Hydrogen burning requires temperatures in excess of $10^{7\circ}$K unless the densities are sufficiently high (see Chapter 8). The existence of a low-mass limit for main-sequence stars arises because the central density increases and the central temperature decreases as the mass is decreased. Stars of sufficiently low mass will be degenerate. The condition for degeneracy is

$$kT < E_F \tag{2-108}$$

where E_F is the electron Fermi energy. The virial theorem requires

$$kT + E_F \sim \frac{Gm_H^2 N}{R} . \tag{2-109}$$

Since $R \propto N^{-1/3}$ for degenerate configurations (see Section 8–3), Equations (2–108) and (2–109) imply

$$kT < E_F \propto N^{4/3}. \tag{2-110}$$

Relation (2–110) shows that a star will no longer be able to maintain nuclear reactions if its mass is sufficiently low. More accurate calculations show that the limit is $\simeq 0.05$ M$\odot$.

2-9   STELLAR WINDS

The prediction and subsequent discovery of a solar wind made it evident that stars similar to the sun should also possess stellar winds. Nonequilibrium processes that arise because of the outer convective zone serve as the energy source that drives the solar wind. Red giants, which possess very extended convective envelopes and large radii (100–500 R$\odot$), are likely candidates for enhanced stellar winds. Recent infrared measurements show that a significant fraction of luminous red giants are surrounded by low-temperature clouds. Some of these infrared objects are radio sources that emit strongly at wavelengths corresponding to transitions of the OH and H_2O molecules. It has been shown that the supergiant α Herculis is losing mass at an appreciable rate. Doppler shifted absorption features that appear to come from an expanding shell of matter are observed in the spectrum of its binary companion, and consequently it is evident that the mass that appears to be flowing from α Herculis exceeds the required escape velocity. Estimates for the rate of mass loss from α Herculis range from 10^{-6}–10^{-9} M$\odot$/yr. The rate of mass loss must approach 10^{-5} M$\odot$/yr for luminous red giants if it is to be significant from the point of view of stellar evolution. Pulsations are a likely mechanism for driving a stellar wind since all luminous red giants are intrinsic variables. Radiation pressure is an additional plausible mechanism.

It is widely believed that most stars initially more massive than the Chandrasekhar mass limit ($\simeq 1.4$ M$\odot$) lose a large fraction of their mass before entering the white dwarf state. Several arguments have been advanced to support this contention. If we assume that star formation is in a steady state within our galaxy and compare the estimated death rate of stars per cubic parsec with the estimated rate of production of white dwarfs within the solar neighborhood, we find that these rates are approximately the same. Supernovae are presumed to arise from evolved stars that are more massive than ~ 1.4 M$\odot$. The relatively low rates of supernovae per galaxy (~ 1 every 70–300 yr) as compared to white dwarfs ($\gtrsim 3$ per year) suggests that most stars are less massive than the Chandrasekhar limit before they enter their final state. The existence of white dwarfs in young galactic clusters provides additional evidence that many massive stars lose sufficient mass to evolve into the white dwarf state.

Another potentially more convincing argument concerning the overall rate of mass loss from stars is obtained by comparing the total amount of energy radiated by main-sequence stars with that from evolved stars (primarily red giants). Although stars consume only ~ 10–15 percent of their hydrogen content on the main sequence, most of the integrated optical luminosity of stars is due to main-sequence stars. This conclusion, which would appear to indicate that most massive stars consume only a fraction of their nuclear energy content before entering the white dwarf state, is strengthened by the result that stable white dwarfs can contain only a small amount of hydrogen. Since luminous red giants emit most of their radiant energy in the infrared, the correctness of the above conclusion will remain in doubt until the total infrared radiation from evolved stars have been determined.

The sun is known to be losing mass at a rate of 10^{-13}–10^{-14} M$\odot$/yr. As we shall show below, the observed high temperature of the solar corona $\sim 2 \times 10^{6\circ}$K implies that the outer region of the solar atmosphere cannot be in hydrostatic equilibrium. For hydrostatic equilibrium to hold, it is necessary that

$$\frac{dP}{dr} = -\rho(r)g(r) \qquad (2\text{--}111)$$

where

$$g(r) = g(a)\left(\frac{a}{r}\right)^2.$$

We assume that a polytropic relation[2] exists between the pressure and density. This assumption implies

$$P(r) = P(a)\left[\frac{\rho(r)}{\rho(a)}\right]^\gamma \qquad (2\text{--}112)$$

where γ is some constant, and $a = 10^7$ km is the radius of a mass shell inside the corona. At $r = a$, it is observed that

$$g(a) = 1.4 \times 10^4 \text{ cm/sec}^2$$
$$n(a) = 10^7 \text{ cm}^{-3}$$
$$T(a) = 2 \times 10^{6\circ}\text{K}.$$

If Equations (2–111) and (2–112) are integrated from the base of the corona to radius r, we find

$$P(r) = P(a)\left[1 - \Lambda(a)\left(1 - \frac{a}{r}\right)\right]^{\gamma/(\gamma-1)} \qquad (2\text{--}113)$$

where

$$\Lambda(a) = \left[\frac{(\gamma-1)}{2\gamma}\right]\frac{m_p g(a)a}{kT(a)}.$$

For ionized hydrogen, the relation between pressure and temperature is

$$P(r) = 2n(r)kT(r). \qquad (2\text{--}114)$$

The pressure must vanish for sufficiently large radii if hydrostatic equilibrium is to hold. This requirement is satisfied if $\Lambda(a) \geq 1$. Since the maximum value

---

[2] See Section 3–8 for a discussion of polytropes.

of γ is $\frac{5}{3}$ which occurs for adiabatic expansion, the condition $\Lambda(a) \geq 1$ implies

$$kT(a) \leq \frac{m_p g(a)a}{5}. \qquad (2\text{-}115)$$

Equation (2-115) shows that the thermal energy at any point in the photosphere must not exceed $\frac{1}{5}$ of the gravitational escape energy $m_p g(a)a$ if the atmosphere is to be in hydrostatic equilibrium. The above-mentioned values for the solar corona show that the solar atmosphere cannot be static.

If $T(r)$ can be measured, then the equation of hydrostatic equilibrium can be integrated, and we find

$$P(r) = P(a) \exp\left[-\frac{m_p}{2k} \int_a^r dr \, \frac{g(r)}{T(r)} \right]. \qquad (2\text{-}116)$$

If $P(r)$ is to approach zero at large radii, the exponent in the above equation must diverge as $r \to \infty$. This leads to the condition that $g(r)/T(r)$ must decrease outwardly less rapidly than $1/r$ for stability. To obtain the steady-state solutions that govern the spherically symmetric outflow from a star, we must integrate the equations of continuity, motion, and energy conservation from some point in the atmosphere where the physical variables are known. For the case of the sun where the mechanisms of energy transport are very complicated and not fully understood, the equation of energy transport is sometimes replaced by an assumed polytropic relation between pressure and density.

For the sun, convection is the ultimate driving mechanism for the mass loss. On the other hand, there exist luminous stars for which radiative energy transport is the likely driving mechanism. The sufficient condition for a stellar atmosphere to be unstable because of radiative energy transport depends on L/M and the opacity κ. We assume that an optically thin, ionized hydrogen layer exists above the surface of a star. The outward force on an electron is

$$F_R = P_R \sigma_T \qquad (2\text{-}117)$$

where $P_R \sim \frac{1}{3}aT^4$, and $\sigma_T = 0.665 \times 10^{-24}$ cm^2 is the Thomson cross section. The gravitational force on a proton is

$$F_G = \frac{Gm_p M}{R^2}. \qquad (2\text{-}118)$$

Electrostatic forces prevent appreciable charge separation. The ionized layer will clearly be unstable if $F_R > F_G$. We use the relation $L = 4\pi R^2 \sigma T^4$ to show that the layer will be unstable if

$$\frac{L}{M} \gtrsim 33{,}000 \, \frac{L\odot}{M\odot}. \qquad (2\text{-}119)$$

The right-hand side of the above equation is multiplied by a factor $(0.4/\kappa)$ if the opacity is a constant value κ. It is important to recognize that we have assumed that the density is too low for convection to be efficient. If the opacity depends on wavelength, the Planck mean opacity should be used for κ if the optical depth is <1. A more careful analysis of the necessary condition for instability would give a somewhat lower required L/M ratio for instability. The above condition for instability may be satisfied for massive stars near the main sequence such as O and B supergiants and Wolf-Rayet stars. These classes of stars show direct evidence for mass loss.

The total time derivative measures how a variable such as the velocity varies in a coordinate system that moves with a particular mass element. For the case of spherical symmetry, it is defined by means of the expression

$$\frac{d}{dt} = \frac{\partial}{\partial t} + v \frac{\partial}{\partial r}. \tag{2–120}$$

The condition that the flow of a gas be in a steady state implies that there is no explicit dependence on time.

If radiation pressure can be neglected, the equations that describe the steady-state spherical expansion of ionized hydrogen are:

The equation of mass conservation

$$r^2 v(r)\rho(r) = r_0^2 v_0 \rho_0 \tag{2–121}$$

where r_0, v_0, and ρ_0 refer to some reference level, and $\rho = nm_p$.

The equation of motion

$$v \frac{\partial v}{\partial r} + \frac{1}{\rho}\frac{dP}{dr} + \frac{GM}{r^2} = 0 \tag{2–122}$$

The energy equation

$$nv\left(3k\frac{dT}{dr} - \frac{2kT}{n}\frac{dn}{dr}\right) = \frac{1}{r^2}\frac{d}{dr}\left(r^2 K(T)\frac{dT}{dr}\right) \tag{2–123}$$

where $K(T) \simeq 6 \times 10^{-7} T^{5/2}$ ergs-sec^{-1}-$^\circ$K^{-1} is the thermal conductivity. Bremsstrahlung losses can be neglected for sufficiently low densities.

The equation of state

$$P = 2nkT. \tag{2–124}$$

Equations (2–121) to (2–124) determine the steady spherical symmetric outflow from a star if values for ρ, P, and v are known at some reference level. The above equations can be integrated for the case of the sun because the values of the physical variables are known in the corona.

If a star is rotating, a stellar wind will cause the loss of angular momentum as well as mass loss. If mass is lost from the surface of an object rotating rigidly out to some radius r, the resultant rate of change of angular momentum is

$$\text{Torque} = \frac{dJ}{dt} = \omega r^2 \frac{dm}{dt} \tag{2-125}$$

where J is the angular momentum of the star; $dm/dt = 4\pi r^2 \rho v$ is the rate of mass loss; and ω is the angular velocity at the surface.

Because magnetic fields can transmit shear, their presence can increase the rate of loss of angular momentum from a star that is losing mass. Therefore, significant surface magnetic fields will cause a region outside the photosphere to corotate with the photosphere and thereby increase the rate of loss of angular momentum. We assume that the magnetic field is primarily radial. The condition that the magnetic flux be conserved in the stellar wind implies that the magnetic field strength falls off inversely as the square of the radius, that is,

$$B = \left(\frac{r_0}{r}\right)^2 B_0 \tag{2-126}$$

where r_0 and B_0 refer to the radius and magnetic field strength, respectively, at the stellar surface.

The "critical radius" for corotation is determined by the condition that the flow velocity v equal the Alfvén velocity, that is,

$$v = v_A \equiv \sqrt{\frac{B^2}{4\pi\rho}} \tag{2-127}$$

and Equations (2-121) to (2-124) which determine how P, ρ, and v vary with r if the influence of the magnetic field on the flow is assumed small. There is evidence that the loss of angular momentum from the sun is larger than would be expected in the absence of surface magnetic fields.

2–10   ROTATION

Although the general features of stellar evolution can be understood by theoretical studies based on nonrotating stars, stellar rotation undoubtedly has an important influence on many stars. This is certainly the case for many massive main-sequence stars whose equatorial rotational velocities are observed to approach the escape velocity. The luminosities of such rapidly rotating stars must be significantly different than nonrotating (spherically symmetric) stars with the same mass and chemical composition.

The equation of hydrostatic equilibrium for a rotating star is

$$\frac{1}{\rho} \nabla P = -\nabla \Phi + \omega^2 \mathbf{R} \tag{2-128}$$

where Φ is the gravitational potential, ω the angular velocity, and $R = r \sin \theta$ the distance from the axis of rotation. It can be shown by means of Equation (2–128) that if the equation of state of a star can be expressed in the form $P = P(\rho)$, the layers of a star in hydrostatic equilibrium must rotate on cylinders (that is, $\omega = \omega(R)$). For such stars the centrifugal force term in Equation (2–128) is derivable from a centrifugal potential. We find

$$V = \Phi - \int_0^R R\omega^2(R) \, dR \qquad (2\text{–}129)$$

where V is the total potential, and the second term on the right-hand side of Equation (2–129) is the centrifugal potential Φ_c. If the star is uniformly rotating, Equation (2–129) becomes

$$V = \Phi - \tfrac{1}{2}r^2\omega^2 \sin^2 \theta. \qquad (2\text{–}130)$$

White dwarfs, neutron stars, polytropes, and convective stars are examples of stars that must rotate on cylinders if they are to remain in hydrostatic equilibrium.

The equation of hydrostatic equilibrium for a star rotating on cylinders is

$$\frac{1}{\rho} \nabla P = -\nabla V. \qquad (2\text{–}131)$$

Equation (2–131) implies that the direction of ∇P and ∇V are parallel. If we take the curl of Equation (2–131), we find

$$\nabla\rho \times \nabla V = 0 \qquad (2\text{–}132)$$

from which it follows that $\nabla\rho$ is parallel to ∇V and ∇P. This circumstance implies that surfaces of constant ρ, P, and V coincide.

If the molecular weight is constant throughout the star and the rotation on cylinders, T as well as P and ρ are constant on surfaces of constant V. Therefore, the radiative flux can be written

$$\mathbf{F} = - \frac{4ac}{3\kappa\rho} T^3 \nabla T = - \frac{4acT^3}{3\kappa\rho} \frac{dT}{dV} \nabla V. \qquad (2\text{–}133)$$

Equation (2–133) shows that the radiative flux is proportional to the effective gravity on a potential surface. Equation (2–133) and the energy equation

$$\nabla \cdot \mathbf{F} = \rho\varepsilon \qquad (2\text{–}134)$$

can be simultaneously satisfied only if

$$\varepsilon \propto \left(1 - \frac{\nabla^2\Phi_c}{4\pi G\rho} \right). \qquad (2\text{–}135)$$

For the case of uniform rotation, Equation (2–135) becomes

$$\varepsilon \propto \left(1 - \frac{\omega^2}{2\pi G\rho}\right). \tag{2–136}$$

Because Equations (2–134) and (2–136) are not likely to be satisfied in stellar interiors, meridional circulation currents, which transport angular momentum as well as heat, must be produced. The velocity of these circulation currents can be shown to be approximately

$$v_c \sim \lambda \left(\frac{L_r}{M_r g(r)}\right) \frac{\bar{\rho}}{\rho(r)} \tag{2–137}$$

where λ is a constant which is approximately equal to the ratio of the rotational energy to the gravitational energy. Equation (2–137), which is not valid right at the surface, predicts that the circulation velocity will be large near the surface. Figure 2–7 shows the predicted paths of circulation currents in the stellar interior. The currents are downward in the equatorial regions where the radiative flux is decreased by rotation and upwards in the polar regions where the radiative flux is increased. Circulation currents will tend to equalize the angular momentum per unit mass along streamlines. If magnetic effects and molecular weight gradients are not important, these circulation currents are expected to redistribute angular momentum in a characteristic time scale,

$$\tau_c \sim \frac{R}{v_c} \tag{2–138}$$

where R is the radius of the star.

The mean angular momentum per unit mass (assuming uniform rotation) is known to be greater for massive main-sequence stars than for stars of relatively low mass such as the sun. The observed mean angular momentum per unit mass is a smooth function of stellar mass except in the neighborhood of ~ 1.5 M$\odot$. Stars more massive than ~ 1.5 M$\odot$ are too hot to possess significant convective envelopes. Such stars are observed to rotate much more rapidly ($v_{surface} \gtrsim 120$ km/sec) than stars of slightly lower mass, which like the sun have significant convective envelopes and undergo very slow rotation ($v_{surface} \sim 2$ km/sec). It is reasonable to assume that low-mass stars lose large amounts of angular momentum during the main-sequence or pre-main-sequence stages. This suggestion is supported by observations. First, the sun is losing angular momentum at a significant rate. Second, observations of relatively young, low-mass main-sequence stars and T Tauri stars suggest much higher rates of mass loss and rotational velocities than presently observed for the sun.

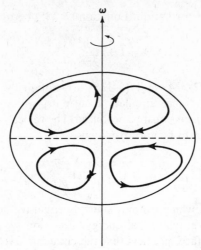

Figure 2–7   The circulation currents of a star rotating with angular velocity ω are shown schematically.

If the sun possessed a rapidly rotating core, then the resulting gravitational quadrupole moment could have an important influence on the expected motion of the perihelion of Mercury. General relativity predicts a perihelion advance of ~ 43 in./century if the sun is a perfect sphere. This prediction is in agreement with experiment. However, recent oblateness measurements of the sun suggest that the sun's quadrupole moment may be sufficient to reduce the predicted advance of the perihelion of Mercury by as much as 4 in./ century. Such a decrease in the predicted advance of Mercury would contradict general relativity.

2–11   GLOBULAR CLUSTERS AND THE AGE OF THE GALAXY

Globular clusters are of great interest because they were probably formed at approximately the same time as the galaxy. For this reason, their ages are a measure of the age of the galaxy, and some of their physical properties, notably their chemical compositions, may be indicative of the physical state of matter at the time the galaxy was formed.

The predicted evolution of a globular cluster star in the Hertzsprung-Russell diagram is shown schematically in Figure 2–8. Hydrogen burning commences in the core of the star at point 1. The path 1–2 represent the main-sequence stage which is dominated by hydrogen core burning. The star spends most of its lifetime ($\sim 10^{10}$ yr) on the main sequence. When hydrogen is exhausted at the center of the star (point 2), a transition from

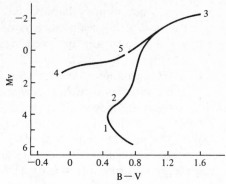

Figure 2-8 A typical globular cluster color-magnitude diagram is shown schematically).

hydrogen core burning to hydrogen shell burning takes place. This first red giant stage is represented by path 2–3. During this phase, which is terminated by the ignition of helium, the main energy source is provided by a hydrogen-burning shell that surrounds an inactive helium core. The mass, degree of degeneracy, and central temperature of the core increase as the star evolves up the red giant branch until the ignition of helium at $L = 2000$–$3000 \, L\odot$ moves the star rapidly (that is, in a Kelvin time) from point 3 to point 5. The loop 4–5, which lasts for $\sim 5 \times 10^7$ yr, represents a phase of helium core and hydrogen shell burning. This predicted evolutionary stage is presently identified with the horizontal branch (see discussion below).

The final evolutionary phase represents the second red giant phase. During this phase the star consists of a degenerate carbon-oxygen core surrounded by a helium-burning shell, a hydrogen-burning shell, and a hydrogen-rich envelope. Stellar model calculations indicate that during this latter stage of evolution, thermal instability is present in the helium-burning shell source. The presence of thermal instability leads to the relaxation oscillations that are discussed in Section 5–4.

An examination of the actual Hertzsprung-Russell diagrams for various globular clusters reveals that important differences exist among different clusters. These observations suggest that heavy-element abundances may have changed appreciably during the epoch when the formation of globular clusters took place.

The study of population II stars (especially globular cluster stars) and the decay of long-lived isotopes are the two basic methods of estimating the age of the galaxy. If the helium abundance, the heavy-element abundance, and the convective mixing length were known, the observed shape of the main-sequence turnoff could be compared with stellar model calculations and an estimate for the age of globular clusters determined. Unfortunately these physical parameters are not securely known, and consequently this method of estimating the ages of globular clusters is not accurate.

A comparison between computed and observed luminosities at main-sequence turnoff is another means by which the ages of globular clusters can be estimated. However, well-known uncertainties in estimating the distances to globular clusters make this method of age determination uncertain. The presence or absence of a post-main-sequence gap in star density on the Hertzsprung-Russell diagram gives an interesting indication of age. The existence of this gap is believed to be related to the presence of a convective hydrogen-burning core which occurs if the temperature-sensitive CNO cycle is the mode of nuclear energy generation. If a star is sufficiently massive to have a convective core during its main-sequence phase, the disappearance of this core at hydrogen exhaustion leads to a rapid phase of evolution that should correspond to an observed gap in the star density on the H-R diagram. The absence of this gap places an upper limit on the masses of stars in globular clusters and therefore a lower limit on their age. This limit is dependent on the assumed chemical composition of the cluster.

Horizontal Branch Stars

Horizontal branch stars have luminosities of $\simeq 50$ L$\odot$ and occupy a region of the H-R diagram that includes the effective temperature range $6000°K \lesssim T_e \lesssim 10^{4°}K$. They are population II stars and are frequently found in globular clusters. Spectroscopic studies indicate that their metallic abundances are typically 10^2 times less than solar abundances. It is widely believed that the post-main-sequence ages of globular clusters are approximately equal to the age of the galaxy. The variable star instability strip intersects the horizontal branch, and consequently variable stars known as RR Lyrae stars occupy a portion of the horizontal branch (see Chapter 4).

Although some important questions regarding the nature of the horizontal branch remain unsettled, stellar evolutionary calculations have provided an explanation for its origin. It is clear that horizontal branch stars represent an evolutionary stage for low-mass (<1 M$\odot$) stars that follows immediately after the ignition of helium (called the helium core flash). During this evolutionary stage, the star is burning helium in its core and hydrogen in a thin shell that surrounds the core. Since the amount of nuclear energy release necessary to lift the degeneracy of the core ($\lesssim 3 \times 10^{49}$ ergs) is much less than the available nuclear energy from helium burning ($\sim 5 \times 10^{50}$ ergs), only a small fraction of the core helium will be consumed during the helium flash.

The nature of the evolution through the horizontal branch depends on the total mass of the star, the initial core mass (that is, the mass inside the hydrogen-burning shell), the helium abundance, and the abundance of heavy elements. Stars more massive than ~ 0.9 M$\odot$ will not evolve toward the blue after the helium flash, and therefore horizontal branch stars must be less massive. However, main-sequence stars in globular clusters may be somewhat more massive since mass loss can occur on the red giant branch prior to the helium flash or perhaps as a result of it.

Although other physical variables such as neutrino emission play a lesser role, the mass of the helium-burning core at the beginning of the horizontal branch is primarily dependent on the initial helium abundance. A high-helium abundance means that the mass of the core at the initial stage of the horizontal branch will be relatively low. This result follows because the onset of helium burning depends primarily on the central temperature which increases as the molecular weight is increased [see Equation (2–94)]. The calculations indicate that a horizontal branch star can evolve from the red to the blue if the hydrogen-burning shell contributes a sufficiently large fraction of the total luminosity. For the case of relatively high initial helium ($Y \gtrsim 0.3$),[3] the star spends much of its time evolving from red to blue. The extent of this evolution to the blue depends sensitively on the initial mass and chemical composition. For low values of Y, the hydrogen shell burning is weaker and the helium burning in the core stronger because the mass of the core is larger. In this case evolution proceeds continuously to the red. Just before the exhaustion of helium at the center, the star evolves rapidly to the red with the subsequent formation of a helium-burning shell. The star ascends the asymptoptic branch during this double shell source phase.

A theoretical argument in support of high-helium abundance in horizontal branch stars is found by comparing the number of horizontal branch stars in a globular cluster with the corresponding number of red giants with greater than horizontal branch luminosity. One assumes that the observed number density of stars is inversely proportional to the lifetime and compares computed lifetimes with observed number densities. When this comparison is made, it is found that the high-helium ($T \gtrsim 0.3$) case is favored over the low-helium ($Y \lesssim 0.1$) case.

It is of considerable interest to compare the properties of horizontal branch stars as inferred from stellar evolutionary calculations with those of RR Lyrae variables as predicted by pulsation studies. The predicted luminosity of the horizontal branch is only slightly higher than expected from pulsation theory and close to that determined by the method of statistical parallaxes. On the other hand, the predicted masses of horizontal branch stars (0.6–0.8 M$\odot$) are somewhat larger than suggested by pulsation theory (0.5–0.6 M$\odot$).

---

[3] Y is the mass fraction of helium.

Decay of Uranium

The observed abundance ratio $[U^{235}/U^{238}]$ can be used to estimate the age of the galaxy. In order to make this estimate, it is necessary to make several assumptions. It is assumed that the observed solar system abundance ratios are representative of the galactic disk at the time of the formation of the solar system and that the abundance ratio $[U^{235}/U^{238}]$ at the time of isotope formation is 1.65, which is what is predicted on the basis of r-process nucleosynthesis. Moreover, it is assumed that a fraction p of long-lived isotopes are formed promptly at the time of the origin of the galaxy and that the remaining fraction are formed at a constant rate.

The rate of formation of a particular isotope on a fractional basis can be written

$$R(t) = p\,\delta(t) + \frac{(1-p)}{T_0} \tag{2-139}$$

where t is set equal to zero at the epoch of galaxy formation, and T_0 is the age of the galaxy. If $R(t)$ is integrated over time from $t = 0$ to $t = T_0$, we find

$$\int_0^{T_0} R(t)\,dt = 1 \tag{2-140}$$

as required by the definition of $R(t)$. If γ is the decay constant of a particular isotope, then its abundance at the time of the formation of the solar system T_s is proportional to

$$N_s = \int_0^{T_s} R(t)e^{-\gamma(T_s-t)}\,dt. \tag{2-141}$$

Integrating Equation (2–141), we find

$$.N_s = \frac{1-p}{\gamma T_0} + \left[p - \frac{(1-p)}{\gamma T_0} \right]\exp\left[-\gamma T_s\right] \tag{2-142}$$

from which the measure of the present abundance

$$N_s \exp\left[-\gamma\,\Delta T\right] \tag{2-143}$$

(where $\Delta T = T_0 - T_s$) can be determined. Since the decay constants of U^{235} and U^{238} are $\gamma_{235} = 0.97 \times 10^{-9}\,\mathrm{yr}^{-1}$ and $\gamma_{238} = 0.154 \times 10^{-9}\,\mathrm{yr}^{-1}$, respectively, Equations (2–141) and (2–143) can be used to obtain an expression for the present abundance ratio $[U^{235}/U^{238}] = 0.00723$ as a function of p, T_0, T_s, and the known decay constants. Since ΔT, the age of the solar system, is known, we can therefore find T_0, the age of the galaxy as a function of p. In principle, the dependence on p can be eliminated by studying the decay of several isotopes. Unfortunately, this procedure has not led to convincing results.

If it is assumed that approximately 50 percent of the heavy-element formation took place promptly after the formation of the galaxy, then the estimated age of the galaxy on the basis of uranium decay is $\sim 7 \times 10^9$ yr. On the other hand, if heavy-element production is assumed continuous, the estimated age of the galaxy is increased to $\sim 15 \times 10^9$ yr. Since the oldest population I stars have heavy-element abundances that are comparable to those of young clusters, there is some observational basis for the prompt formation of long-lived isotopes if it is assumed that they are formed at the same rate as other heavy elements. It is expected that the age of the galaxy should be less than the Hubble time ($\sim 13 \times 10^9$ yr) but greater than the age of the solar system ($\sim 4.6 \times 10^9$ yr). The above estimates for the age of the galaxy are not inconsistent with these upper and lower age limits.

CHAPTER 3

Star Formation and Protostars

3-1 INTRODUCTION

The birth of stars is one of the most fundamental problems in astrophysics. The existence of life is dependent upon physical conditions that arise during and/or immediately after the formation of stars. Because the sun is a very normal star, it would seem plausible that life developed in the neighborhood of other stars. This point of view has been supported by the discovery that some relatively nearby stars have companions of planetary mass (for example, Barnard's star). Moreover, the recent discovery that rather complicated molecules are common in dense interstellar clouds that may be protostars has demonstrated that the molecular components necessary for the development of life pervade interstellar space. Unfortunately there is at the present time no well-developed theory for the formation of stars, and therefore our discussion will be very preliminary. After describing evidence for the recent birth of stars, we shall review some of the basic physical ideas that describe how dense interstellar clouds (protostars) may form from the interstellar medium and then describe how these protostars are likely to evolve into normal stars.

The existence of very luminous and therefore massive main-sequence stars is perhaps the most powerful argument for their recent formation. Such stars must be young because main-sequence lifetime decreases sharply as a function of increasing mass. Stellar evolutionary calculations indicate that stars leave the main sequence after they have exhausted 10 to 15 percent of their initial hydrogen. Since the nuclear energy content of hydrogen is $\simeq 10^{52}$ ergs/M$\odot$, the lifetime of a star on the main sequence is

$$\tau \simeq 10^{10} \frac{M}{\text{M}\odot} \frac{\text{L}\odot}{L} \text{ yr.} \qquad (3-1)$$

Equation (3–1) and the known mass-luminosity relation [see Equation (2–16)] indicates that the main-sequence lifetimes of stars more massive than about 1.5 M$\odot$ is $\simeq 5 \times 10^9$ yr. This result implies that more massive stars must be much younger than the galaxy ($\tau_{\text{galaxy}} \sim 10^{10}$ yr).

A second argument in favor of recent star formation is the strong correlation in interstellar space between the positions of hot stars and interstellar clouds. Both hot stars and gas are concentrated close to the galactic plane. Observational evidence indicates that the rate of star formation in the neighborhood of the sun is proportional to ρ_{gas}^x where $x \simeq 2$. Moreover, it is clear that the spiral structure of the galaxy plays a fundamental role in initiating the formation of stars. Many faint stars including those of the T Tauri type are seen within interstellar clouds. T Tauri stars are above the main sequence on the H-R diagram and presumably contracting toward it. The observed distribution of T Tauri stars within the Orion nebula, which is the closest large HII region, indicates that star formation is widespread. The estimated mean gas density within the Orion nebula is $10^{-22} \pm 1$ g/cm^3. However, there are individual clouds in Orion whose densities exceed 10^{-18} g/cm^3.

Most stars appear to be formed in groups (that is, clusters or associations). The mass of a typical group of stars is 10^2–10^4 M$\odot$. It has been estimated that the number of early main-sequence stars in clusters and associations accounts for approximately all the stars formed in the solar neighborhood. The existence of stellar associations, which are gravitationally unbound, constitutes an important argument in favor of recent star formation. Stellar associations are observed to be expanding and may have been originally formed under conditions of much higher density. It is generally believed that star formation takes place in clusters or associations because the instability that led to the initial collapse of the interstellar cloud required that the cloud have a large mass. Fragmentation of the collapsing interstellar cloud is assumed to take place although a satisfactory physical model describing this fragmentation is lacking.

3–2 VIRIAL THEOREM

We consider a finite, self-gravitating system such that the influence of viscosity and turbulence can be neglected. The equation of motion of an inviscid fluid can be written

$$\rho \frac{dv_i}{dt} = \rho \left(\frac{\partial v_i}{\partial t} + v_j \frac{\partial v_i}{\partial x_j} \right)$$

$$= -\rho \frac{\partial \Phi}{\partial x_i} - \frac{\partial}{\partial x_i} \left(P + \frac{B^2}{8\pi} \right) + \frac{1}{4\pi} \frac{\partial}{\partial x_j} B_i B_j \qquad \left(\begin{matrix} i = 1, 2, 3 \\ j = 1, 2, 3 \end{matrix} \right)$$

$$(3\text{–}2)$$

where Φ is the gravitational potential, and v_i is the velocity in the i direction. If we multiply Equation (3–2) by x_i, integrate over the region, and then integrate by parts, we obtain

$$\frac{1}{2}\frac{d^2I}{dt^2} = 2T + 3\int P\,d^3r + U_{\text{mag}} + \Omega \qquad (3\text{–}3)$$

where $\mathbf{r} = (x, y, z)$ is the position vector,

$$I \equiv \int r^2\rho\,d^3r \qquad (3\text{–}4)$$

T is the kinetic energy of mass motion, such as rotational energy,

$$U_{\text{mag}} = \frac{1}{8\pi}\int B^2\,d^3r \qquad (3\text{–}5)$$

is the magnetic energy, and Ω is the gravitational energy. For a nondegenerate, ideal gas with constant γ, we can write

$$3\int P\,d^3r = 3(\gamma - 1)U \qquad (3\text{–}6)$$

where $\gamma = c_P/c_v$, and U is the internal energy.

Let us assume that the configuration is in equilibrium and that the surface integrals and rotational energy can be neglected. In this case Equation (3–3) reduces to

$$3(\gamma - 1)U + U_{\text{mag}} + \Omega = 0. \qquad (3\text{–}7)$$

That a relation of the form of Equation (3–7) should exist under conditions of equilibrium can be readily understood by means of dimensional analysis. At equilibrium the gas pressure and magnetic pressure must balance the gravitational forces. It follows that

$$P_{\text{gas}} + P_{\text{mag}} = P_{\text{grav}} \qquad (3\text{–}8)$$

where

$$P_{\text{gas}} = C_1\frac{U}{V}$$

$$P_{\text{mag}} = \frac{B^2}{8\pi} = C_2\frac{U_{\text{mag}}}{V}$$

and

$$P_{\text{grav}} = \rho g R = -C_3\frac{\Omega}{V}.$$

Equation (3–8) has a form similar to Equation (3–7), except for the fact that the C's are not determined in Equation (3–8).

If a configuration is in equilibrium and surface terms can be neglected, the total energy is

$$\mathscr{E} = U + U_{\text{mag}} + \Omega. \tag{3-9}$$

Equation (3-7) can be used to eliminate U from Equation (3-9). We obtain

$$\mathscr{E} = -\frac{(3\gamma - 4)}{3(\gamma - 1)}(|\Omega| - U_{\text{mag}}). \tag{3-10}$$

A configuration will become dynamically unstable if $\mathscr{E}$ is positive. Equation (3-10) shows that no equilibrium configurations will exist if

$$U_{\text{mag}} > |\Omega| \tag{3-11a}$$

or

$$(3\gamma - 4) < 0. \tag{3-11b}$$

The first condition in Equation (3-11) implies that if $\gamma > \frac{4}{3}$, the configuration will be unstable when

$$|\Omega| < \frac{1}{8\pi}\int B^2 \, d^3x = \tfrac{1}{6}R^3\langle B^2\rangle \tag{3-12}$$

where

$$\Omega = -\frac{3}{5}\frac{GM^2}{R}$$

for a gas sphere of uniform density. Equations (3-10), (3-11), and (3-12) imply that for a fixed magnetic field, a critical mass M_c is required if a configuration with $\gamma > \frac{4}{3}$ is to be gravitationally bound. If the density of the cloud is constant and the effect of gas pressure can be neglected, the critical mass becomes

$$M_c \gtrsim \frac{(5/2G)^{3/2}(B/\rho^{2/3})^3}{48\pi^2}. \tag{3-13}$$

Typical interstellar conditions are $B \sim 3 \times 10^{-6}$ G and $\rho \sim 20m_p$ cm^{-3}. It follows that

$$M_c \sim 10^3\text{--}10^4 \text{ M}\odot. \tag{3-14}$$

This result demonstrates that magnetic fields cannot be ignored in star formation. The critical mass given above would not be reduced by quasi-spherical collapse since the dependence of the gravitational energy and magnetic energy on the radius of the configuration is the same. The gravitational energy per unit volume is

$$\rho\,\frac{GM}{R} \propto \rho^{4/3} M^{2/3}. \tag{3-15}$$

The conductivity of the gas is sufficiently high that the magnetic flux is probably conserved during collapse so that

$$B \propto \frac{1}{R^2} \propto \rho^{2/3}. \tag{3-16}$$

Equations (3–5), (3–15), and (3–16) show that the ratio of magnetic and gravitational energies remain unchanged during quasispherical collapse. From Equations (3–13) and (3–16), it follows that M_c remains unchanged during quasispherical collapse, and consequently masses as low as 1 M⊙ are unlikely to form in this manner. However, in the presence of an appreciable magnetic field, the motion of the gas would likely be along the magnetic lines of force, and consequently the critical mass can be reduced.

Expression (3–11b) implies that a configuration will be dynamically unstable if $\gamma < \frac{4}{3}$. The dissociation of molecular hydrogen will reduce the value of γ below $\frac{4}{3}$. It has been suggested and more recently shown by means of detailed calculations that the dissociation of molecular hydrogen can cause dynamical instability during the collapse of a protostar. Dynamical instability of this general type may also play a role in the origin of supernovae (see Chapter 9).

If angular momentum is conserved during collapse, centrifugal forces will prevent overall collapse in two dimensions while turbulence and thermal pressure very likely set a lower limit to the masses that can be formed by the flattening of the cloud. If ω_0 is the initial angular velocity of the gas cloud, the conservation of angular momentum implies

$$\omega_0 r_0^2 = \omega r^2 \qquad (\theta \sim \text{constant}). \tag{3-17}$$

At any point (r, θ, ϕ) the ratio of centrifugal to gravitational force is

$$\frac{\omega^2 r \sin \theta}{(GM_r/r^2) \sin \theta} = \left[\frac{\omega_0^2 r_0^4}{GM_r} \right] \frac{1}{r} \tag{3-18}$$

and consequently the centrifugal force will eventually dominate gravitational force during collapse.

In the above discussion we have argued that the presence of magnetic fields and rotation will make quasispherical collapse unlikely. However, it has been shown that if pressure can be neglected, a very small asymmetry will grow rapidly during collapse (that is, a slightly oblate spheroid will quickly become a flat disk while a slightly prolate object will quickly collapse into a cigar-shaped object even in the absence of rotation or appreciable magnetic forces. This result indicates that spherical symmetry cannot be maintained during a pressure-free collapse.

So far we have ignored the effect of external pressure on the collapse of a gas cloud. Because an isothermal gas sphere has a finite density at all finite radii, the total mass inside the sphere would approach ∞ as $r \to \infty$. This means that an isothermal gas sphere will be in equilibrium only if acted on by a finite external pressure P_{ext}.

We can obtain an approximate solution for an equilibrium configuration by means of the virial theorem. The appropriate form of the virial theorem is

$$4\pi P_{ext} R^3 = 3(\gamma - 1)U + \Omega \qquad (3-19)$$

where

$$U \simeq \frac{3kTM}{2\mu m_H} = \frac{3}{2} M v_s^2$$

$$\Omega \simeq -\frac{3}{5}\frac{GM^2}{R}$$

μ = molecular weight

v_s = isothermal sound velocity

$$\gamma \simeq \frac{5}{3}.$$

Therefore, Equation (3-19) becomes

$$4\pi P_{ext} R^3 = 3M v_s^2 \left[1 - \frac{\frac{3}{5}(GM^2/R)}{3M v_s^2} \right]. \qquad (3-20)$$

For an equilibrium to exist in the absence of external pressure, the right-hand side of Equation (3-20) must be negative, which means that

$$R < \lambda_J \sim \frac{v_s}{\sqrt{G\bar{\rho}}} \qquad (3-21)$$

where λ_J is the Jeans length. Equation (3-20) shows that for a fixed temperature and value of M, there exists a maximum external pressure for which equilibrium configurations are possible. The critical radius (that is, λ_J) is reached when

$$\frac{\partial P_{ext}}{\partial R} = 0. \qquad (3-22)$$

The corresponding mass is $\sim \rho \lambda_J^3$, that is,

$$M_c \sim \frac{v_s^2}{G^{3/2}\rho^{1/2}}. \qquad (3-23)$$

For $T = 100°K$ and $\rho/m_H = 20$ cm^{-3}, which are typical values for dense interstellar clouds, $M_c \sim 10^4$ M$\odot$. This value for M_c is about the mass of a galactic cluster. It is interesting to note that for primeval clouds for which we might guess $T \sim 10^{4°}K$ and $\rho/m_H \sim 1$ cm^{-3} (see Chapter 13), $M_c \sim 1 \times 10^6$ M$\odot$. This is about the characteristic mass of a dwarf galaxy.

3–3 DYNAMICAL INSTABILITY OF INTERSTELLAR GAS

We assume that an infinite medium is initially at rest with uniform density ρ_0, pressure P_0, and gravitational potential Φ_0. Although these assumptions are mutually inconsistent since

$$\nabla^2 \Phi_0 = 4\pi G \rho_0 \qquad (3\text{–}24)$$

exact calculations for finite systems give similar results, and therefore the condition for instability obtained in this manner is of considerable interest.

The linearized equations of continuity, motion, Poisson's equation, and the equation of state are obtained by substituting $\rho = \rho_0 + \rho_1$, $P = P_0 + P_1$, $\Phi = \Phi_0 + \Phi_1$, and $v = v_1$, and collecting terms of the same order. The linearized equations become

$$\frac{\partial \rho_1}{\partial t} + \rho_0 \nabla \cdot \mathbf{v}_1 = 0 \qquad (3\text{–}25)$$

$$\rho_0 \frac{d\mathbf{v}_1}{dt} = -\nabla P_1 - \rho_0 \nabla \Phi_1 \qquad (3\text{–}26)$$

$$\nabla^2 \Phi_1 = 4\pi G \rho_1 \qquad (3\text{–}27)$$

$$P_1 = \rho_1 \frac{kT}{\mu m_H} = \rho v_s^2 \qquad (3\text{–}28)$$

where v_s is the isothermal sound velocity. We assume that the medium remains isothermal under compression and search for solutions of the form

$$\rho_1 = A e^{i(kx - st)}. \qquad (3\text{–}29)$$

Taking the gradient of the second equation, we find

$$s^2 = k^2 v_s^2 - 4\pi G \rho_0. \qquad (3\text{–}30)$$

The plane wave solution will be unstable if $s^2 \leq 0$, that is, if the wavelength of the plane wave solution is greater than the Jeans length

$$\lambda_J = \sqrt{\frac{\pi kT}{\mu m_H G \rho_0}}. \qquad (3\text{–}31)$$

The physical basis for Jeans criterion for gravitational instability can be understood if we compare the release of energy per gram by means of a pressure wave ($\sim v_s^2$) with the simultaneous reduction in gravitational energy ($\sim G\rho \lambda_J^2$). When the Jeans criterion is satisfied, the configuration is unstable because it can reach states of lower energy by means of small amplitude perturbations.

During the early stages of collapse, the temperature inside a gas cloud is expected to remain approximately constant. This circumstance implies that the Jeans length defined in Equation (3–31) will become smaller as the cloud collapses, and consequently smaller clouds inside the main cloud may themselves become unstable to collapse. It has been suggested that this process of fragmentation will continue until the fragmented clouds become optically thick to their own radiation at which point the increased internal temperature and pressure caused by the trapping of the radiation will prevent further fragmentation.

In our discussion above, we have ignored the presence of the galactic magnetic field, which is dynamically coupled to the interstellar gas. It can be shown that the condition for the hydrostatic equilibrium of the interstellar medium given in Equation (1–141) is unstable. The time scale for the growth of the instability is comparable to the free-fall time. This instability arises because a slow compression of the interstellar gas does not appreciably change its temperature, and consequently the gas does not resist compression. Any perturbation that compresses the magnetic lines of force at one point and expands them at an adjacent point disturbs the equilibrium in such a manner that the interstellar gas moves along the magnetic lines of force into the compressed region. The additional gas that is thus added to the compressed region tends to compress the region further, and so the initial perturbation can develop to large amplitude. In a similar manner, the motion of the gas decreases the weight of the gas in the expanded region, and therefore it can expand further.

The Jeans instability, which arises because of self-gravitation, requires rather large masses ($\sim 10^4$ M$\odot$) as we have noted above. On the other hand, it can be shown (see Figure 3–1) that the presence of an interstellar magnetic field reduces the effective gravity of the composite medium by a factor

$$\sim \left(1 + \frac{g^2}{GB^2} \right)$$

and so smaller mass clouds can become unstable.

3–4   LOSS OF ANGULAR MOMENTUM

It has been pointed out above that angular momentum must be lost from a collapsing interstellar gas cloud if star formation is to take place. Magnetic fields can transport angular momentum, and therefore it is of interest to consider a simple model that illustrates how a magnetic field can transport angular momentum from a gas cloud.

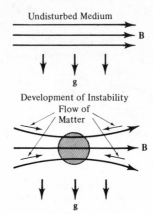

Figure 3–1   The development of a dynamical instability in the presence of a magnetic field is shown schematically.

We are given an axially symmetric gas sphere of radius R_0 with rotational and magnetic axes parallel. The magnetic field is assumed to extend into the interstellar medium as shown in Figure 3–2. At time $t = 0$ the cloud is made to rotate with angular velocity ω. The surrounding interstellar medium of density ρ is assumed stationary at $t = 0$. The equations that describe the motions outside the gas sphere are

$$E = -\frac{v}{c} \times B \tag{3-32}$$

$$\frac{1}{c}\frac{\partial B}{\partial t} = \nabla \times \left(\frac{v}{c} \times B\right) \tag{3-33}$$

$$\rho\frac{dv}{dt} = \frac{1}{4\pi}(\nabla \times B) \times B \tag{3-34}$$

$$\nabla \cdot B = 0. \tag{3-35}$$

The first of the above equations follows from the high conductivity of the gas [see Equation (1–124)]. The assumed axial symmetry and the further assumptions that $v_R = v_z = B_R = 0$ for all t allow us to reduce the above equations to the following

$$\dot{B}_\theta = B(R)\frac{\partial v_\theta}{\partial z} \quad (B_z \equiv B(R)) \tag{3-36}$$

$$\dot{v}_\theta = \frac{B(R)}{4\pi\rho}\frac{\partial B_\theta}{\partial z} \tag{3-37}$$

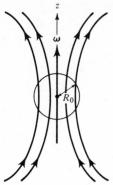

Figure 3–2   A gas cloud of radius R_0 is shown rotating with angular velocity ω in a magnetic field.

where R, θ, and z are the usual cylindrical coordinates. Equations (3–36) and (3–37) give

$$\ddot{v}_\theta - \frac{B^2}{4\pi\rho}\frac{\partial^2 v_\theta}{\partial z^2} = 0 \tag{3–38}$$

$$\ddot{B}_\theta - \frac{B^2}{4\pi\rho}\frac{\partial^2 B_\theta}{\partial z^2} = 0 \tag{3–39}$$

where

$$v_A = \sqrt{\frac{B^2}{4\pi\rho}}$$

is the Alfvén velocity. Both v_θ and B_θ satisfy the wave equation, and therefore their solutions follow immediately. For $z > v_A t$,

$$\begin{aligned} v_\theta &= 0 \\ B_\theta &= 0. \end{aligned} \tag{3–40}$$

For $z < v_A t$ and positive z direction,

$$v_\theta = v\left(t - \frac{z}{v_A}, R\right)$$

$$B_\theta = -(4\pi\rho)^{1/2} v\left(t - \frac{z}{v_A}, R\right). \tag{3–41}$$

The above equations give the following rate of transfer of angular momentum due to Alfvén waves propagating in the positive z direction:

$$I(t) = 2\pi \int_0^{R_0} v\left(t - \frac{z}{v_A}, R\right) v_A \rho R^2 \, dR \tag{3–42}$$

where R_0 is the radius of the rotating cloud, and ρ the density of the interstellar medium. At the surface of the cloud, the velocity is

$$v(t, R) = \omega(t)R. \tag{3-43}$$

Therefore,

$$I(t) = \frac{\pi^{1/2}}{4} \rho^{1/2} B R_0{}^4 \omega(t). \tag{3-44}$$

If J is the angular momentum of the gas cloud ($J \sim \frac{2}{5}MR_0{}^2\omega$), we find

$$\frac{dJ}{dt} + 2I = 0. \tag{3-45}$$

The factor 2 arises because waves can propagate in either the $\pm z$ direction. Substituting into the above equation gives

$$\frac{2}{5}MR_0{}^2 \frac{d\omega}{dt} + \frac{2\pi^{1/2}}{4} \rho^{1/2} B R_0{}^4 \omega = 0. \tag{3-46}$$

Solving this equation, we find

$$\tau = \frac{4}{5} \frac{M \ln (\omega_i/\omega_f)}{\pi^{1/2}\rho^{1/2}BR_0{}^2} \tag{3-47}$$

where τ is the characteristic time for the surface of the cloud to slow down from $v_i = \omega_i R_0$ to $v_f = \omega_f R_0$. The slowdown arises because as the Alfvén waves propagate into the interstellar medium, they cause an increasingly large mass to rotate.

3-5 LOSS OF MAGNETIC FLUX

Although a neutral gas can move freely through a magnetic field, ions and electrons are tied to the magnetic field as a result of the high conductivity and the large characteristic length of the gas cloud. Even in relatively dense HI clouds, about 1 atom in 10^3 are probably ions. Soft x rays and low-energy cosmic-ray particles are expected to be the principal source of ionization in these clouds since UV radiation from stars cannot penetrate. We wish to determine the time scale for neutral matter to separate from the magnetic field if an applied force such as a pressure gradient is pushing the neutral gas across the magnetic lines of force. The frictional drag on the neutral atoms must balance the applied force when the limiting drift velocity is attained. At the limiting velocity, the frictional force per unit volume is[1]

$$\mathbf{F} = n_i n_H \langle \sigma_{H_i} u \rangle m_p (\mathbf{v} - \mathbf{v}_i) \tag{3-48}$$

---

[1] Rate coefficient $\equiv \langle \sigma u \rangle$ is average over assumed Maxwell-Boltzmann distribution.

where u is the thermal velocity, v the mean velocity of the neutral gas, and v_i the mean velocity of the ions. The applied force becomes $-kT\nabla n_H$ if it is due to a pressure gradient and the gas is isothermal. Equating the applied force and the frictional force, it follows that

$$|\mathbf{v} - \mathbf{v}_i| = \left| \frac{kT\nabla n_H}{m_p n_i \langle \sigma_{H_i} u \rangle n_H} \right| \simeq \left| \frac{kT}{m_p n_i \sigma_{H_i} u L} \right| . \tag{3-49}$$

Since $\sigma_{H_i} \sim 10^{-15}$ cm^2, $u \sim 10^5$ cm/sec, and $n_i \sim 10^{-3}$ cm^{-3}, the time scale for neutral gas to diffuse a distance L is

$$\tau_{\text{Diff}} \sim \frac{L}{|v - v_i|} \sim \frac{L^2}{10^{22}} \text{ sec.} \tag{3-50}$$

This diffusion time scale is 10^9–10^{10} years for 10 pc clouds and, therefore, is longer than the corresponding free-fall time. If the applied force arises from the bending of the magnetic lines of force, it can be written

$$F = \frac{(\nabla \times \mathbf{B}) \times \mathbf{B}}{4\pi}$$

$$\sim \frac{B^2}{8\pi L} . \tag{3-51}$$

Therefore, $|\mathbf{v} - \mathbf{v}_i|$ becomes

$$|\mathbf{v} - \mathbf{v}_i| = \frac{B^2}{m_p n_i \langle \sigma_{H_i} u \rangle n_H 8\pi L} . \tag{3-52}$$

The diffusion time can be shortened somewhat by neutralization of ions on grains. Moreover, it is possible that more complicated plasma effects can lead to the enhanced diffusion of the neutral gas across the magnetic lines of force.

3–6 SPHERICALLY SYMMETRIC COLLAPSE OF PROTOSTAR

The simplest model for star formation is a spherically symmetric collapsing gas cloud of constant density. It is assumed that at $t = 0$, $v_0(M_r) = 0$ and that the pressure remains negligible during collapse. M_r is the mass interior to radius r.

The equation of motion for each mass shell can be written

$$\frac{d^2 r}{dt^2} = -\frac{GM_r}{r^2}, \qquad r = r(M_r) \tag{3-53}$$

$v = dr/dt$ can be regarded as a function of r or M_r. Therefore,

$$\frac{dv}{dt} = \frac{dv}{dr}\frac{dr}{dt} = \frac{1}{2}\frac{dv^2}{dr}$$

$$\frac{1}{2}\frac{dv^2}{dr} = -\frac{GM_r}{r^2} \tag{3-54}$$

$$\tfrac{1}{2}[v^2 - v_0^2(M_r)] = GM_r\left[\frac{1}{r} - \frac{1}{r_0(M_r)}\right].$$

$v_0(M_r)$ and $r_0(M_r)$ are the velocities and positions at $t = 0$ of the mass shell with mass M_r interior to it.

This equation can be integrated analytically:

$$\left(\frac{2GM_r}{r_0}\right)^{1/2} t = (r_0(M_r) - r)^{1/2}r^{1/2} + r_0(M_r)\sin^{-1}\left(\frac{r_0 - r}{r_0}\right)^{1/2}. \tag{3-55}$$

For $\rho_0(M_r) = \text{constant} = \rho_0$, $M_r = (4\pi r_0^3(M_r)/3)\rho_0$, we find

$$\left(\frac{8\pi G\rho_0}{3}\right)^{1/2} t = \left(1 - \frac{r}{r_0}\right)\left(\frac{r}{r_0}\right)^{1/2} + \sin^{-1}\left(1 - \frac{r}{r_0}\right)^{1/2}. \tag{3-56}$$

That Equations (3–55) or (3–56) are solutions of Equation (3–54) can be checked by differentiation. Equation (3–56) implies that if the initial density is uniform, the mass shells will reach the origin (that is, $r = 0$) at the same time (τ_{ff}).

$$\tau_{ff} = \left(\frac{32G\rho_0}{3\pi}\right)^{-1/2} \quad \text{(free-fall time)}. \tag{3-57}$$

The above solution is not very realistic for several reasons. (1) It ignores the effect of the boundary conditions since the density has been assumed constant everywhere (that is, even at infinity). If the central density is initially greater than the mean density, then it can be shown that the central shells will fall proportionally faster than the outer shells. Under such conditions a shock wave will form near the center of the star during collapse. (2) The assumption that the pressure is negligible must eventually become invalid as the cloud collapses. Moreover, it can be shown that if the pressure is strictly zero, small nonradial perturbations will make the collapse deviate from spherical symmetry by a large factor. With suitable small perturbations, it will collapse to a pancake-shaped object (that is, become very oblate) or to a prolate. (3) The effects of magnetic fields and angular momentum are likely to play a crucial role in the collapse of a gas cloud.

The early evolution of a protostar is characterized by three basic time scales. The free-fall time has been discussed above [Equation (3–57)]. The expansion time scale is derived under the assumption that gravity can be neglected. In this case the equation of motion becomes

$$\frac{d^2r}{dt^2} = \frac{1}{2}\frac{dv^2}{dr} \sim \frac{1}{2}\frac{d}{dr}\left(\frac{r}{t}\right)^2 \sim \frac{1}{\rho}\frac{dP}{dr} \tag{3-58}$$

and therefore means of dimensional analysis we find

$$\tau_e \sim \sqrt{\frac{\rho}{P}}\, r = \left(\frac{\mu m_H}{kT}\right)^{1/2}\left(\frac{3M}{4\pi\rho}\right)^{1/3} \tag{3-59}$$

where M is the mass of a protostar, and μ is the molecular weight. The Kelvin time scale, which is associated with the heating or cooling of the isothermal gas cloud, is

$$\tau_K = \frac{3k\rho T}{2\mu m_H(\Lambda - \Gamma)} \tag{3-60}$$

where Λ is the cooling rate per unit volume, and Γ is the heating rate per unit volume. The relative magnitudes of Λ and Γ determine whether the cloud is heated or cooled. If the free-fall time is less than the Kelvin time scale, the star is able to adjust itself with sufficient rapidity to follow changes in its state produced by energy losses or gains and thus can reach hydrostatic equilibrium. This condition is not a sufficient one since the protostar might be dynamically unstable. On the other hand, if the Kelvin time is shorter than the free-fall time, the star cannot be in hydrostatic equilibrium since it is unable to follow small changes in its temperature.

If a protostar is transparent to its own radiation, Kirchhoff's law implies that its luminosity is

$$L = \int j_v \, dv \, d\Omega \, d^3r = \int k_v I_v \, dv \, d\Omega \, d^3r \tag{3-61}$$

where k_v is the monochromatic absorption coefficient, and $d\Omega$ is an element of solid angle. If I_v can be replaced by the blackbody function $B_v(T)$, the luminosity becomes

$$L = M4\pi\bar{\kappa}\sigma T^4 \tag{3-62}$$

where $\bar{\kappa}$ is the Planck mean opacity.[2] If the protostar is not transparent to its own radiation, the diffusion equation is appropriate for computing the luminosity and thus

$$L = -4\pi r^2 \frac{4ac}{3}\frac{T^3}{\kappa\rho}\frac{dT}{dr} \sim -\frac{50\sigma M^{1/3}T^4}{\kappa\rho^{4/3}} \tag{3-63}$$

---

[2] The opacity κ is related to the absorption coefficient k by the expression $\kappa\rho = k$.

where κ is the Rosseland mean opacity (see Section 2–5). The final relation in the above equation was obtained by approximating $(dT/dr) \sim (T/R)$ and $R \sim (3M/4\pi\rho)^{1/3}$. Figure 3–3 shows schematically the general behavior of a collapsing gas cloud in the $\rho - T$ plane. The values for ρ and T denote central density and temperature, respectively. Because the opacity is very uncertain, it is not clear when the protostar becomes opaque to its own radiation. Interstellar grains are a likely source of opacity. The opacity due to grains has been estimated to be ~ 0.1 for near infrared radiation. If this estimate is correct, protostars will become opaque when $0.1\rho R \sim 1$. For a 1 M$\odot$ star this means

$$\kappa\rho \left(\frac{M}{\rho}\right)^{1/3} \sim 1$$

$$(3\text{–}64)$$

or

$$\rho \sim 10^{-16} \text{ g/cm}^3.$$

There are a number of physical processes that determine the rates of heating and cooling of a collapsing protostar during its transparent stage. Heating is probably caused by low-energy cosmic ray particles and soft x rays. Since interstellar clouds are nearly transparent to cosmic ray particles,

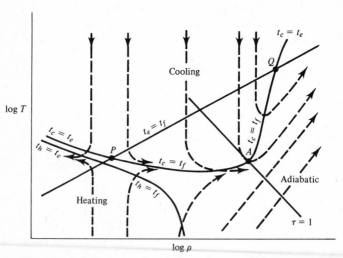

Figure 3–3 The schematic features of evolution of a star of given mass are shown in the log T-log ρ diagram. The star is opaque in regions above the curve $\tau = 1$. The dashed curves represent the evolutionary tracks. The time-scales of cooling, heating, expansion, and free fall are denoted by t_c, t_h, t_e, and t_f, respectively.

Source: From Hayashi, C. 1966, *Ann. Rev. Astron. Ap.*, **4**, 171, by permission of Annual Reviews, Inc.

direct heating due to these particles could be readily found if their fluxes were known. Excitation of Alfvén waves by means of cosmic rays is another mechanism for heating the interstellar gas. It is plausible, although by no means certain, that soft x rays produced during supernova explosions and by other galactic x-ray sources play an important role in ionization and heating. Cooling is caused primarily as a result of collisional excitation of excited states of heavy ions such as C^+, Si^+, and Fe^+ by electrons and collisional excitation of the fine structure components of C and O atoms by H atoms. Collisions between atoms and grains tend to maintain the gas and grains at constant temperature. Collisional excitation of the rotational levels of molecules such as CO and H_2 is an additional important cooling mechanism.

If it is assumed that most heavy elements in interstellar space reside in grains, the number of grains per cubic centimeter can be estimated since the radii of grains are $\sim 10^{-5}$ cm. This implies that $n_g = 10^{-13} - 10^{-14}n$ where $n = \rho/m_p$. The temperature of grains in interstellar medium has been estimated to be $\sim 20°K$. This temperature is determined by the energy gain due to absorption of starlight and collisions with the surrounding gas and cooling due to reradiation in the infrared. It is important to emphasize that, in general, the temperature of the surrounding gas T will not be the same as the grains T_g. If the gas temperature is higher than that of the grains, the grains will act to cool the gas and conversely. We assume that if a H atom or H_2 molecule collides with a grain and becomes attached to the surface, it will remain in contact with the grain for a sufficiently long time that it will exchange an amount of energy $k(T - T_g)$ with the grain before being ejected. The expression for the cooling (or heating) by grains can be written

$$\Lambda_g = n_g[f_H n_H \langle v \rangle_H + f_{H_2} n_{H_2} \langle v \rangle_{H_2}] \pi r_g^2 k(T - T_g) \qquad (3\text{-}65)$$

where $\langle v \rangle = (8kT/\pi m)^{1/2}$ and f_H (and f_{H_2}) denote the fraction of collisions for which the H atom or H_2 molecule becomes trapped on the surface of the grain.

The equation that determines the energy balance between grains, starlight, and interstellar gas is

$$\Lambda_g + \bar{\kappa}\rho_g I = \bar{\kappa}(T_g)\rho_g \frac{ac}{4} T_g^4$$

$$\begin{pmatrix} \text{collisions} \\ \text{with gas} \end{pmatrix} + \begin{pmatrix} \text{absorption of} \\ \text{starlight} \end{pmatrix} = (\text{emission in I.R.}) \qquad (3\text{-}66)$$

where $\bar{\kappa}(T_g)$ is the blackbody mean opacity for grains at temperature T_g; ρ_g is the mean density of matter in grains; and I is the intensity of background radiation.

Molecular hydrogen is expected to be very abundant in dense interstellar clouds where it is protected from UV radiation. The absence of appreciable

21-cm radiation from some dense interstellar clouds suggests that hydrogen is primarily in the molecular state. For temperatures $< 150°K$, hydrogen molecules are in the ground rotational level, $J = 0$. The principal excitation is with H atoms to the $J = 2$ states which are $E = 0.044$ eV above the $J = 0$ state. The radiative lifetime of the $J = 2$ state is $A_{20}^{-1} = 2.4 \times 10^{-11}$ sec. Under steady-state conditions, we can write

$$n_0(n_{H_2}\gamma_{02} + n_H\gamma'_{02}) = n_2(A_{20} + n_{H_2}\gamma_{20} + n_H\gamma'_{20}) \qquad (3\text{-}67)$$

where the γ's are the rate coefficients

$$\gamma_{ij} = \langle \sigma_{ij}v \rangle = \frac{4}{\pi^{1/2}} \left(\frac{m_r}{2kT}\right)^{3/2} \int_0^\infty v^3\sigma_{ij}(v)e^{-(mv^2/2kT)} \, dv \qquad (3\text{-}68)$$

m_r is the reduced mass, and n_0 and n_2 are the number density of H_2 molecules in the $J = 0$ and $J = 2$ states, respectively. From the principle of detailed balance, we find

$$\gamma_{02} = \gamma_{20} \frac{g_2}{g_0} e^{-(E/kT)}. \qquad (3\text{-}69)$$

A similar relation holds for γ'_{02} and γ'_{20}. The cooling rate becomes

$$\Lambda_{H_2} = A_{21}En_2$$

$$\simeq A_{21}En_{H_2} \frac{g_2}{g_0} e^{-(E/kT)} \frac{(n_{H_2}\gamma_{20} + n_H\gamma'_{20})}{(A_{20} + n_{H_2}\gamma_{20} + n_H\gamma'_{20})} \qquad (3\text{-}70)$$

where we have assumed that $e^{-(E/kT)}$ is small.

There is observational evidence that the CO molecule is abundant in dense interstellar clouds. Because the CO molecule is much more easily excited at low temperatures than the H_2 molecule; it is likely to be a much more effective cooling agent than H_2.

3-7   ENERGETICS OF COLLAPSING GAS CLOUD (SIMPLE MODEL)

A spherically symmetric gas cloud with initial radius R_0 and constant density ρ_0 is assumed to collapse uniformly (that is, $\rho \sim 1/R^3$). The gas cloud is assumed to remain isothermal until radius $R = R_1$ and then to become adiabatic for $R < R_1$. The gravitational energy release in collapsing from R_0 to R_1 is

$$\Omega = -\frac{3}{5} \frac{GM^2}{R_1} \left[1 - \frac{R_1}{R_0}\right]. \qquad (3\text{-}71)$$

The heat generated by isothermal compression and radiated away is

$$-M \int_{\rho_0}^{\rho_1} P \, d\left(\frac{1}{\rho}\right) = M v_s^2 \ln\left(\frac{\rho_1}{\rho_0}\right) = \frac{3MkT}{\mu m_H} \ln\left(\frac{R_0}{R_1}\right). \quad (3\text{-}72)$$

If $R_1 \ll R_0$ the gravitational energy release is much greater than the radiative energy losses. This means that most of the gravitational energy release is converted into kinetic energy of infall.

We have assumed that compressional heating is ineffective in raising the temperature of a transparent, collapsing gas cloud. Such an increase in temperature could halt collapse. The assumption of approximately isothermal collapse is correct because the cooling rate increases more rapidly with increasing density than the compressional heating rate.

During adiabatic collapse between R_1 and R_2 ($R_2 < R_1$), the increase in thermal energy is

$$U_2 - U_1 = -M \int_{\rho_1}^{\rho_2} P \, d\left(\frac{1}{\rho}\right)$$

$$= \frac{M}{(\gamma - 1)}\left[\frac{P_2}{\rho_2} - \frac{P_1}{\rho_1}\right] = U_1\left[\left(\frac{R_1}{R_2}\right)^{3(\gamma-1)} - 1\right]. \quad (3\text{-}73)$$

The above equation shows that if $\gamma < \frac{4}{3}$ as is the case when H_2 dissociates, the rate of increase of thermal energy as a function of radius will be less rapid than the gravitational energy release, and consequently collapse will continue. On the other hand, if $\gamma > \frac{4}{3}$, thermal energy will increase more rapidly than gravitational energy, and the collapse will be stopped. If the collapse is stopped at $R = R_2$ where $R_2 \ll R_0$, energy conservation implies

$$\Omega \approx -U_2. \quad (3\text{-}74)$$

The above expression indicates that if the simplifying assumptions made in constructing this model are to hold, a protostar will have an excess amount of thermal energy when collapse is stopped, since the virial theorem requires

$$3(\gamma - 1)U_2 + \Omega = 0 \quad (3\text{-}75)$$

for equilibrium and zero surface pressure. Hydrodynamic calculations of collapsing gas clouds show that collapse is halted in the central regions. A shock wave separates the core with the infalling outer envelope, which supplies the external pressure necessary to maintain the core in quasihydrostatic equilibrium.

There exists a minimum surface temperature for stars that are in hydrostatic equilibrium. Although the precise value for this minimum temperature depends on the opacity, convection theory and the luminosity of the star, the existence of a minimum surface temperature can be readily understood. If the radius of a star becomes sufficiently large, the outer layers become optically thin, and consequently radiative transfer is very rapid. The resulting low temperatures and correspondingly low pressures are not sufficient to maintain the extended envelope in hydrostatic equilibrium, since the characteristic Kelvin time for the outer layers becomes shorter than its free-fall time. Because protostars are not in hydrostatic equilibrium, they can and are expected to have very low surface temperatures.

3–8 POLYTROPES

A self-gravitating configuration is called a polytrope of index n if the relation

$$P = K\rho^{1+1/n} \qquad [K = K(R, M)] \tag{3–76}$$

is valid throughout. Such polytropic configurations have well-known properties that follow from Poisson's equation and the requirement of hydrostatic equilibrium.

For the case of spherical symmetry, Poisson's equation is

$$\frac{d^2\Phi}{dr^2} + \frac{2}{r}\frac{d\Phi}{dr} = -4\pi G\rho \tag{3–77}$$

when $-\Phi$ is the gravitational potential as defined in other sections of the book. Hydrostatic equilibrium implies

$$dP = -g\rho \, dr = \rho \, d\Phi. \tag{3–78}$$

If a polytropic relation between pressure and density is assumed, Equation (3–78) becomes

$$\left(1 + \frac{1}{n}\right) K\rho^{1/n-1} \, d\rho = +d\Phi. \tag{3–79}$$

Integrating Equation (3–79), we find

$$(n + 1)K\rho^{1/n} = +\Phi + \text{constant.} \tag{3–80}$$

If we choose $\Phi = 0$ at $\rho = 0$, it follows that

$$\rho = \left[\frac{\Phi}{(n + 1)K}\right]^n$$

$$P = \frac{\rho\Phi}{n + 1}. \tag{3–81}$$

Inserting the above expression for ρ into Equation (3–77), we find the Emden equation:

$$\frac{d^2\Phi}{dr^2} + \frac{2}{r}\frac{d\Phi}{dr} + \alpha^2\Phi^n = 0$$

$$\alpha^2 = \frac{4\pi G}{[K(n + 1)]^n}. \tag{3–82}$$

The Emden equation can be solved numerically subject to the additional boundary conditions: $\Phi = \Phi_c$ and $d\Phi/dr = 0$ at $r = 0$. The solutions to the Emden equation have. been tabulated for various polytropic indices. The computed ratio of central density ρ_c to mean density increases with assumed polytropic index. This ratio is 5.99 for $n = \frac{3}{2}$ and becomes infinite at $n = 5$.

3–9 FULLY CONVECTIVE STARS

The transition stage from collapse to hydrostatic equilibrium is a stage in which the kinetic energy of mass motion is converted into thermal energy. It has been proposed that at the end of the transition stage, the distribution of entropy S throughout the star approaches that of quasihydrostatic, convective equilibrium (that is, $S \sim$ constant) throughout the star.

Fully convective stars in quasihydrostatic equilibrium are expected to be approximately polytropes with $n = \frac{3}{2}$ because the relation

$$\frac{\rho}{P}\left(\frac{dP}{d\rho}\right)_s = \gamma = 1 + \frac{1}{n} \tag{3-83}$$

with $\gamma = \frac{5}{3}$ (except in hydrogen and helium ionization zones and regions where H_2 dissociates) holds throughout. This is the case because the super-adiabatic temperature gradients necessary to transport the convective energy flux are small unless the luminosity is very high.

Since the thermal and gravitational energies of the initial stage of adiabatic collapse are likely to be small as compared to the initial stage of quasihydro-static equilibrium, we can obtain a rough estimate of the mass-radius relation for a fully convective pre-main-sequence star if we set the energy of ionization (H, H_2, and He) equal to $\frac{1}{2}$ the gravitational energy,

$$\frac{1}{2}\left(\frac{3}{5-n}\right)\frac{GM^2}{R} = X\frac{M}{m_p} \tag{3-84}$$

where $n = \frac{3}{2}$ and $X = 15.8$ eV is the ionization and dissociation energy per proton for a pure hydrogen star. The above equation implies

$$\frac{R}{R\odot} \approx 50\frac{M}{M\odot} \tag{3-85}$$

where

$$\rho_c = \frac{5.99M}{(4\pi/3)R^3}$$

for a polytrope of index $n = \frac{3}{2}$. If the mass and radius are given, the central temperature of a polytrope $n = \frac{3}{2}$ can be determined from the expression

$$\frac{kT_c}{\mu m_p} = 0.538\frac{GM}{R}. \tag{3-86}$$

For a pre-main-sequence star $T_c \sim 1.5 \times 10^{5\circ}$K, which is much lower than required to support nuclear energy generation, and so gravitational energy release will supply the luminosity. The remaining values for P, ρ, T, and r as a function of mass fraction can be estimated from the polytropic relation between pressure and density and the equation of state for an ideal gas.

The luminosity of a fully convective star can be estimated by determining the amount of energy radiated from the star's photosphere. The tenuous outermost layers of a star must be in radiative equilibrium because the convective velocities cannot exceed the local speed of sound. If we assume that the photosphere is in radiative equilibrium and use the Eddington approximation (see Appendix B), we find

$$\tau_p = \tfrac{2}{3}$$

$$P_p = \frac{Ag_p}{\kappa_p} \tag{3-87}$$

$$g_p = \frac{GM}{R^2}$$

where p denotes photosphere, and $A = \tfrac{2}{3}$ if the opacity is constant. In order to determine the luminosity, we assume a central temperature T_c, and then from Equations (2-70) and (3-76), it follows that

$$P = K(R, M)T^{2.5} \qquad (n = \tfrac{3}{2}). \tag{3-88}$$

To find the effective temperature, which is the temperature at which $P = P_p$, we assume that Equation (3-88) holds throughout the star. Since the radius is known once the mass, polytropic index, and central pressure are specified, the luminosity,

$$L = 4\pi R^2 \sigma T_e^{\;4} \tag{3-89}$$

is determined.

Since the central temperature is too low for nuclear energy generation, contraction is the only available energy source, and we have

$$L \simeq \frac{d(GM^2/R)}{dt}. \tag{3-90}$$

From Equations (3-86) and (3-90) it follows that the central temperature of the star increases as it contracts toward the main sequence. The luminosity of the contracting star decreases primarily because the opacity increases in its outermost layers.

The opacity at the center of the star decreases as its central temperature increases since it is primarily Kramers opacity ($\kappa \propto \rho T^{-7/2}$). Eventually, the central radiative temperature gradient becomes less than the adiabatic temperature gradient, and a radiative core develops. At this stage, the star is approaching a luminosity minimum. The luminosity increases after the opacity in the radiative core, which now includes a large fraction of the mass of the star, becomes sufficiently low. The star reaches the main sequence when its central temperature becomes sufficiently high ($\gtrsim 10^{7\circ}$K) to sustain nuclear energy generation.

3–10 ENTROPY PRODUCTION AND THE TRANSITION TO HYDROSTATIC EQUILIBRIUM

The entropy per gram of a nonrelativistic monatomic gas is

$$S = c_v \ln \left[\frac{T^{3/2}}{\rho} \right] + \text{constant} \qquad (3\text{–}91)$$

where c_v is the heat capacity per gram. The entropy of a contracting star must decrease in the absence of nuclear sources. It follows that the entropy per gram, which is approximately constant throughout a convective star, is a monotonic function of its luminosity. The entropy necessary to produce pre-main-sequence stars is generally greater than that of interstellar clouds. For this reason, entropy production must take place during collapse if a protostar is to evolve into a fully convective star. It has been suggested that the shock wave (or waves) that are produced during collapse will increase the entropy by an amount sufficient to make this transition possible. Turbulence can also cause the entropy to be increased. However, because supersonic turbulence is highly dissipative, turbulence is not likely to become very important after the protostar has become opaque to its own radiation.

We limit our discussion to spherically symmetric collapse for which realistic hydrodynamic calculations are possible. The collapse of a dense gas cloud commences after the Jean's condition [see Equation (3–31)] is satisfied. Hydrodynamic calculations show that if the initial gas cloud is only slightly centrally condensed, the inner regions collapse more rapidly than the outer layers. This effect is self-accelerating, and consequently the central temperature and density increase rapidly as compared to the outer layers. At a temperature of about 1300°K, the dissociation of the H_2 molecule decreases the effective value of $\gamma = c_p/c_v$ below $\frac{4}{3}$, and consequently the protostar becomes dynamically unstable. The diversion of thermal energy into energy of dissociation and later ionization energy of hydrogen slows down the rate of increase of the central temperature. Eventually, the force due to gas pressure exceeds that due to gravity near the center of the protostar, and consequently collapse is halted within this region. Since the outer mass shells are falling supersonically with respect to the decelerated core, a shock wave forms outside the core. Entropy production takes place across this shock front. Numerical calculations indicate that the entropy production across the shock front is sufficient to produce convective pre-main-sequence stars for protostars of moderate mass. For more massive protostars ($M \gtrsim$ 50–100 M⊙), the predicted entropy production is insufficient.

In Section 2–8 it was shown that radiation pressure is the dominant source of pressure in very massive stars. It was suggested that the observed scarcity of such stars may be related to physical conditions surrounding their formation. The entropy per gram necessary to form very massive stars increases sharply as a function of mass, primarily because of the increased entropy

$$\frac{4}{3}\frac{aT^3}{\rho}$$

due to radiation. Since interstellar gas clouds begin with approximately the same entropy per gram, the required entropy production during collapse increases as a function of stellar mass. Hydrodynamic calculations indicate that for the case of spherically symmetric collapse, protostars more massive than 100–200 M⊙ are unable to produce enough entropy during collapse to form a main-sequence star of the same mass. This result implies that very massive, chemically homogeneous stars cannot form directly from spherical collapse. If the collapse times are sufficiently long as compared to the predicted pre-main-sequence lifetimes ($\sim 10^4$ yr), then the collapse of a very massive protostar will lead to the formation of a less massive main-sequence star that accretes matter from the collapsing gas cloud. It has been suggested that after this main-sequence star attains some critical mass, which is a function of its luminosity and the opacity of the dust in the surrounding gas cloud, its emitted radiation is able to prevent further collapse. On the other hand, if the protostar collapse time is sufficiently short, then the nuclear sources will turn on in a shell. Such collapse may lead to unstable behavior. It should be emphasized that rotation and other more complicated physical processes than those we have described are likely to play a fundamental role in determining the stellar mass spectrum.

3–11 OBSERVATIONAL EVIDENCE

1. T Tauri stars appear to be convective stars that are contracting toward the main sequence and occur only in association with nebulosities that contain hot, young stars. They have strong emission lines and show considerable infrared excesses. Mass loss is apparently taking place, since absorption lines are seen on top of emission lines.

2. Infrared objects that have surface temperatures too low to maintain hydrostatic equilibrium are found. Although most of these objects are probably associated with the final stages of stellar evolution, some are likely to be connected with the early stages.

3. Dark nebulae are easily recognizable on photographs of stars, and consequently they have been known to astronomers for many years. Some of these dark nebulae are dense, round objects called globules that are seen projected against diffuse nebulae. The spatial distribution of globules is similar to that of B stars, and their linear dimensions are $\sim 20,000$ AU. Estimates of their density and temperatures would allow us to determine whether or not they are gravitationally bound and, therefore, likely candidates for protostars. Their estimated densities ($n_H \sim 10^5$–10^6 cm^{-3}) indicate that globules would be gravitationally bound if their temperatures were $\lesssim 10°$K.

4. The presence of molecules in interstellar space (see Section 1–9) is a powerful means of studying the physical properties of interstellar clouds. For example, the discovery of certain HCN transitions indicates that some interstellar clouds have densities as high as 10^6 atoms/cm^3. Such clouds are probably collapsing since their temperatures must be low.

CHAPTER 4
Variable Stars

4–1 OBSERVATIONS

A star is said to be pulsationally unstable if it is unstable against oscillations of small amplitude. These small oscillations will grow until they reach some limiting amplitude that is determined by nonlinear effects. Those variable stars, which are intrinsic, periodic variables, are believed to be pulsationally unstable. Many variable stars pulsate with a single periodicity. There are, however, exceptions; for example, some β canis majoris variables. In addition to its great intrinsic interest, the study of variable stars is related to several astronomical problems of fundamental importance, notably the extragalactic distance scale, the initial galactic helium abundance, and the age of the galaxy. The approximate positions of variable stars in the H-R diagram are shown in Figure 4–1. RR Lyrae stars and Cepheids are found exclusively on a narrow instability strip in the H-R diagram.

In order to develop a theory of variable stars, it is necessary to solve the equations of motion, energy conservation, and energy transfer and then compare the results (that is, periods, luminosities, radii, amplitudes of light curves, phase shifts, shapes of light curves) with observations. In general, the mass and chemical composition are the physical variables that can be used to fit the observations.

Classical Cepheids. They are young, relatively massive stars (5–10 $M\odot$) that have high luminosity ($\sim 10^4$ $L\odot$) and, therefore, can be observed in external galaxies. A phase lag between the time of minimum radius and that of maximum luminosity is characteristic of pulsating stars that are on the variable star instability strip, as is the rapid rise and relatively slow decline of the light curve. For Cepheid variables, the Doppler shifts of the spectral lines indicate that the star is brightest when it is expanding through its time

116

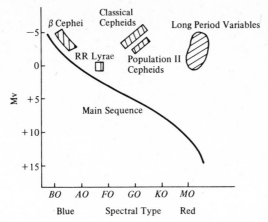

Figure 4–1   The regions of the H–R diagram that are occupied by the most important classes of periodic variable stars are shown schematically.

averaged radius, and consequently there is an approximately 90° phase lag
between maximum luminosity and minimum radius (see Figure 4–2).

The theory of variable star pulsation explains the characteristic phase
shifts between their luminosity and velocity curves. The phase shift that is
observed for variable stars within the instability strip is caused by ionization
zones that have sufficient heat capacity to delay the flow of energy. A crude

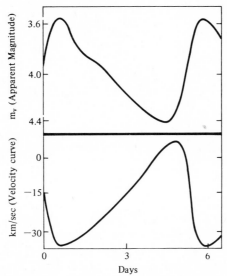

Figure 4–2   A typical Cepheid, light curve and velocity curve are shown
schematically.

approximation for the phase shift between the luminosity and velocity curves is

$$\varphi = \frac{2\pi}{PL} \int c_v T \, dM \tag{4-1}$$

where P is the pulsation period, L the average luminosity, and the integration is over the mass external to the ionization zones.

The existence of a narrow instability strip means that a well-defined period-luminosity-color relation will exist for stars of similar mass. The empirical period-luminosity relations for classical Cepheids and RR Lyrae variables are used to establish the extragalactic distance scale.

Population II Cepheids.   In addition to classical Cepheids, which are massive stars, there exist population II Cepheids called W Virginis stars. Recent stellar evolutionary calculations have indicated that certain helium shell-burning stars of low-mass (0.65 M⊙) and high-helium content (35 percent by mass) will pass through a loop in the H-R diagram that reaches into the strip occupied by population II Cepheids. Therefore, it appears that the occurrence of such loops might explain the origin of some population II Cepheids.

RR Lyrae Variables.   They are horizontal branch stars and are observed in globular clusters. The amplitude of their light variations is ~ 1 magnitude. Their masses are estimated to be quite low (0.5–0.7 M⊙ on the basis of nonlinear pulsation calculations). Their observed heavy-element abundances are generally much less than solar abundances, and it is believed that these low values are representative of those existing prior to the formation of the galaxy. On the other hand, pulsation studies predict that the helium abundance of these objects is relatively high (20–30 percent by mass). This result is of great interest because, in its simplest form, the big bang theory for the origin of the universe predicts that the amount of helium produced during the fireball stage should be approximately this value.

β Canis Majoris Variables (*β* Cephei variables).   They show small amplitude (≤ 0.1 mag) light variations throughout a cycle and therefore are hard to detect. This means that the total number of these variables is probably much larger than the number observed. Their observed surface temperatures (8000–16,000°K) and estimated luminosities suggest that they occupy a region of the H-R diagram that is 1–2 magnitudes above the upper main sequence. The pulsations of β canis majoris variables are generally stable, and there is no evidence of an expanding envelope. The observed sharp absorption lines (H, He, C^+) indicate low rotational velocities (≤ 60 km/sec). This may be an observational effect since it is easier to observe the spectral variations of weakly varying stars if they are slow rotators. At the present time the physical mechanism for the origin of β canis majoris variables is not understood. Their observed velocity-luminosity curves, which show that maximum luminosity occurs at minimum radius, as well as their position on the H-R diagram indicate that ionization zones play little or no role in causing the pulsations. Some β canis majoris variables are doubly periodic with the two periods very nearly the same. For β canis majoris, these periods are $6^h 00^m$ and $6^h 02^m$. Stellar calculations indicate that the double periodicity observed for some β canis majoris stars is likely to be the result of a degeneracy between the frequency of the fundamental radial oscillatory mode and a nonradial (Kelvin) mode of a 10–15 M$\odot$ star that is 1–2 magnitudes above the main sequence. It is possible that nearly all sufficiently massive stars pass through a β canis majoris phase as they evolve off the main sequence.

Long Period Variables (Mira variables). They are very red variables that emit most of their radiation in the infrared. Their luminosities are quite high ($\sim 10^4$ L$\odot$), and their characteristic time scales for variability very long (~ 6 months). The visual magnitudes of these stars vary by 1–10 magnitudes between maximum and minimum. Their infrared spectra are dominated by H_2O absorption, and their visual spectra are characterized by TiO bands. They are found in both populations I and II. At least one Mira variable has been observed in a globular cluster (47 Tucanae). It is believed that most stars of low and intermediate mass pass through a Mira variable stage near the tip of the red giant branch and therefore probably just before the star passes through the planetary nebula stage of stellar evolution. It has recently been discovered that many Mira variables are OH and H_2O radio sources, and it is likely that significant mass loss takes place during this stellar evolutionary stage.

4–2 PULSATION PERIOD

The fundamental radial pulsation period of a star is approximately

$$\text{period} \simeq \frac{2R}{v_s} \tag{4-2}$$

where $v_s = \sqrt{\gamma P/\rho}$ is the speed of sound, and R the radius of the star. Since the energy in pulsations is much less than the gravitational energy of the star, the virial theorem gives

$$-\Omega \simeq \frac{GM^2}{R} \simeq 3 \left\langle \frac{v_s^2}{\gamma} \right\rangle M. \tag{4-3}$$

Equations (4-2) and (4-3) imply

$$\text{period} \simeq \left(\frac{I}{-\Omega}\right)^{1/2} \simeq \left(\frac{1}{G\bar{\rho}}\right)^{1/2} \tag{4-4}$$

where

$$I = \int_0^M r^2 \, dM_r$$

$$M = \tfrac{4}{3}\pi R^3 \bar{\rho}$$

and the right-hand side of Equation (4-4) is the characteristic free-fall time. Observations as well as more precise calculations for pulsating stars within the Cepheid-RR Lyrae instability strip shown on Figure 4-1 indicate that

$$\text{period} \simeq 0.022 \, \frac{(R/R\odot)^{1.7}}{(M/M\odot)^{0.7}} \, \text{days} \tag{4-5}$$

is more nearly correct. The pulsation periods for stars range from $\lesssim 10^{-3}$ sec for neutron stars to $\gtrsim 10^3$ days for some long period variables.

4-3   CRITERION FOR PULSATIONAL INSTABILITY

The stellar energy equation can be written

$$\frac{dQ}{dt} = \varepsilon - \frac{dL_r}{dM_r}. \tag{4-6}$$

If contractional energy release can be neglected, the right-hand side of Equation (4-6) is equal to zero in the equilibrium state, and so the first-order energy equation becomes

$$\frac{dQ}{dt} = \delta\varepsilon - \frac{d\,\delta L_r}{dM_r}. \tag{4-7}$$

If a steady pulsation is to be maintained, the internal energy U of each mass element must remain constant from cycle to cycle. From the first law of thermodynamics [Equation (2-69)] and the constancy of U from cycle to cycle, it follows that

$$W = \oint \int \frac{dQ}{dt} \, dM_r \, dt \tag{4-8}$$

where W is the total work done if we sum over all mass elements. For steady pulsations, the entropy is also a perfect differential,

$$dS = \frac{1}{T} \frac{dQ}{dt} \, dt \tag{4-9}$$

and therefore constant from cycle to cycle. Integrating over a complete cycle, we find

$$0 = \oint dS = \oint \frac{1}{T} \frac{dQ}{dt} \, dt \qquad (4\text{-}10)$$

for each mass element. Since variations in T are assumed small, we can write

$$\frac{1}{T} \simeq \frac{1}{T_0} - \frac{\delta T}{T_0^{\,2}} \qquad (4\text{-}11)$$

where T_0 is the temperature of the undisturbed state. Equations (4-10) and (4-11) imply

$$\oint \int \frac{dQ}{dt} \, dM_r \, dt = \oint \int \frac{\delta T}{T_0} \frac{dQ}{dt} \, dM_r \, dt \qquad (4\text{-}12)$$

and therefore the average rate of change of oscillatory energy is

$$\left\langle \frac{dW}{dt} \right\rangle = \left\langle \int \frac{\delta T}{T_0} \left(\delta\varepsilon - \frac{d \, \delta L_r}{dM_r} \right) dM_r \right\rangle \qquad (4\text{-}13)$$

where the brackets in Equation (4-13) denote time average. When the above integral is positive (negative), the star is pulsationally unstable (stable). The kinetic energy of pulsations is

$$J = \left(\frac{2\pi}{\text{period}} \right)^2 \int \left(\frac{\delta r}{r} \right)^2 r^2 \, dM_r. \qquad (4\text{-}14)$$

The reciprocal e-folding time[1] for the buildup of pulsational energy is

$$\frac{\langle dW/dt \rangle}{2 \langle J \rangle}$$

where the factor of 2 arises because, on the average, one half the pulsational energy is kinetic, and one half is potential. Considerations of dimensional analysis show that this e-folding time is a Kelvin time scale and therefore usually much longer than the pulsation period which is a dynamical time scale. However, these time scales become comparable for luminous red giants (see Section 5-5).

Equation (4-14) makes it clear that there are two basic mechanisms for exciting pulsations. Most stars will be stable with respect to excitation by means of nuclear sources primarily because the relative amplitude of the oscillatory modes are generally very low in the mass zones occupied by the nuclear sources. However, pulsational instability caused by nuclear energy

---

[1] The e-folding time is the time for the pulsational energy to grow by a factor of $e \equiv 2.71828$.

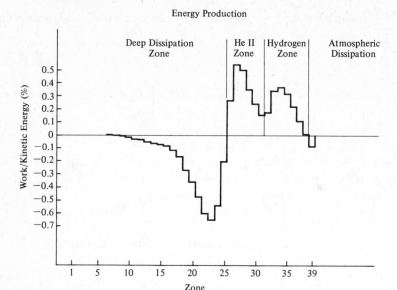

Figure 4-3   The relative amounts of energy production and dissipation are shown for a number of mass shells of an RR Lyrae variable.

Source: From Christy, R. 1966, *Ap. J.*, **144**, 108, by permission of the University of Chicago Press.

generation may be of considerable importance with regard to the theory of novae, x-ray sources (Chapter 7), and massive stars. The second mechanism for exciting pulsations, which is called the valve mechanism, depends on the properties of the He^+ and H ionization zones. To test for pulsational instability, one finds the modes of oscillation and then uses these modes to determine the stability integral [Equation (4–13)]. The adiabatic modes can be used to evaluate the stability integral so long as the mass shells of greatest importance are such that $\tau_K > \tau_D$, where τ_K is the time for a photon to diffuse from a particular mass shell to the surface. For cases where the H ionization zone is important, the adiabatic approximation is no longer adequate, and nonadiabatic modes must be used.

It follows from the second law of thermodynamics that the valve mechanism for exciting pulsations is effective (that is, luminous energy is converted into kinetic energy) if excess heat is trapped during the compression stage and then released during the expansion stage. It is clear that the opacity κ will help excite pulsations if it is higher during the compression stage and lower during the expansion stage than the corresponding equilibrium values. However, if Kramers' opacity ($\kappa \propto \rho T^{-7/2}$) dominates, the opacity decreases rapidly with temperature, and opacity effects will usually tend to stabilize a star. However, in an ionization zone where matter is partially

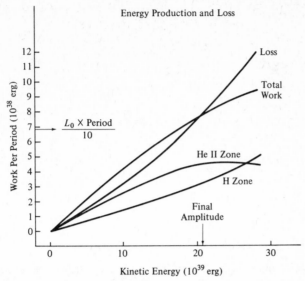

Figure 4-4   The energy production and loss per period are shown as a function of the kinetic energy of pulsation.

Source: From Christy, R. 1966, *Ap. J.*, **144**, 108, by permission of the University of Chicago Press.

ionized, γ is close to unity. Since $P\rho^{-\gamma}$ = constant during an adiabatic compression, little change in temperature occurs throughout an oscillation, and consequently the energy of compression goes into ionizing the gas. For this reason, the opacity can help excite pulsations. However, it would be possible to excite pulsations even if the opacity were independent of ρ and T. This destabilizing effect, which is called the Γ effect, can arise if the ionization zone is favorably situated [that is, $\varphi \sim 1$ in Equation (4-1)]. A mass shell that is partially ionized is relatively cooler than its surroundings during compression and thus will absorb excess heat. This excess heat serves to increase the pressure during the subsequent expansion and consequently provides an extra push that tends to build up pulsations. Figure 4-3 shows how the various mass shells contribute toward the buildup of pulsations for a RR Lyrae variable.

The pulsational amplitude reaches a maximum when the total work done during a period is balanced by the dissipation. As shown in Figure 4-4, the limiting pulsational amplitude arises because of nonlinear behavior associated with the He II (He$^+$) ionization zone.

CHAPTER 5
Red
Giants

5-1 INTRODUCTION

If the apparent magnitude of stars in a single cluster or the absolute magnitudes of stars with known distances are plotted as a function of color, two
major classes of stars are most prominent: main-sequence stars and red giants.
Main-sequence stars are known to be nearly homogeneous stars with varying
mass that are burning hydrogen in their cores. Red giants are stars that have
evolved off the main sequence and, therefore, are chemically inhomogeneous.
Hydrogen burning takes place in a thin hydrogen shell that surrounds a
hydrogen exhausted core. Red giants are characterized by extensive convective
envelopes, very large radii ($R \gtrsim 100 R\odot$), and high luminosities. Although
extensive numerical calculations are necessary to fully explain the structure
and physical behavior of red giant stars, physical arguments can make the
results of these calculations plausible. Such arguments, as well as a further
description of the properties of red giants, will be presented in the subsequent
sections of this chapter. The structure of a red giant is shown in Figure 5–1.

Since they represent the major intermediate evolutionary stage between
the main sequence and the final state of a star, red giants play a fundamental
role in the theory of stellar evolution. They are sites for nucleosynthesis and
extensive mass loss. Moreover, the origin of planetary nebulae and most
supernovae is likely to be a direct consequence of the final evolution of red
giants.

5-2 CONVECTIVE ENVELOPES

Main-sequence stars with spectral type later than about F2 (that is, $M \lesssim 1.5$
$M\odot$), as well as red giants, have significant outer convective envelopes. The
presence of such convective envelopes is a consequence of the temperature
and density dependence of opacity and also the reduction of the adiabatic

gradient that is caused by the partial ionization of H and He. Although convective envelopes may extend nearly to the surface, the outermost layers of a star must be in radiative equilibrium. This is the case because convective velocities cannot exceed the local speed of sound and consequently, as the density becomes sufficiently low, convective energy transport becomes inefficient.

If the Eddington relation between temperature and optical depth is assumed to hold in the radiative layer near the surface of a star, we have (see Appendix B)

$$T^4 = \tfrac{1}{2}T_e{}^4(1 + \tfrac{3}{2}\tau) \qquad (5\text{--}1)$$

where T_e is the effective temperature, and τ is the optical depth. Equation (5–1) implies

$$\frac{dT}{d\tau} = \frac{3T}{8 + 12\tau}. \qquad (5\text{--}2)$$

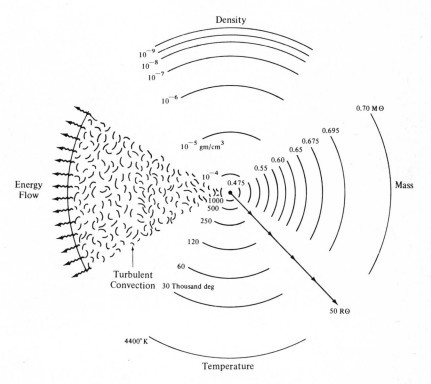

(a) Red Giant Envelope

Figure 5–1   Some important physical properties of the interior of a red giant are shown schematically. (Based on calculations by Iben.)

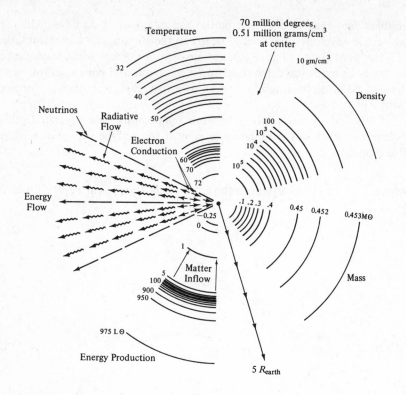

(b) Core of Red Giant

Figure 5–1 *(continued)*

In order for convection to occur, the radiative gradient must exceed the adiabatic gradient, which is

$$\left(\frac{dT}{d\tau}\right)_{ad} = \frac{\gamma - 1}{\gamma}\frac{T}{P}\frac{dP}{d\tau}.$$ (5–3)

The quantity $T^{\gamma}P^{1-\gamma}$ is constant during the adiabatic motion of a convective element. The condition that the surface layer be in hydrostatic equilibrium implies

$$\frac{dP}{d\tau} = \frac{g}{\kappa}.$$ (5–4)

If the opacity κ does not vary with τ, we find immediately from Equation (5–4) that

$$\frac{1}{P}\frac{dP}{d\tau} = \frac{1}{\tau}.$$ (5–5)

Consequently, it follows from Equations (5-2), (5-3), and (5-5) that the condition for convection is

$$\frac{\gamma - 1}{\gamma} \leq \frac{3\tau}{8 + 12\tau}. \tag{5-6}$$

The inequality (5-6) shows that if γ is $\geq \frac{4}{3}$, convection will not take place for any value of τ if the opacity is constant. However, in Late-type (cool) stars the opacity tends to decrease sharply as the surface is approached. When H^- is the dominant opacity source, as is the case for stars like the sun as well as most red giants, the opacity is proportional to the free electron density which decreases sharply with temperature. It can readily be shown by means of the above equations that if the opacity increases with increasing pressure (for example, $\kappa \propto P^n$ with $n \geq 1$), convection commences at relatively small optical depths.

Although the precise color of a red giant depends on the physical nature of the opacity law (that is, on whether it is caused by H^- or H_2O, and so on), physical uncertainties associated with the theory of convection are at least of equal importance in determining the theoretical position of the red giant branch. Moreover, it should be emphasized that even if the opacity were constant, very red stars could not exist in hydrostatic equilibrium because at sufficiently low densities, the Kelvin times associated with the outer layers would become less than the corresponding free-fall times.

5-3 SCHÖNBERG-CHANDRASEKHAR LIMIT

In the above discussion we have outlined the physical requirements for the existence of convective envelopes in red giants and Late-type main-sequence stars. However, we have not explained why stars should evolve to the red giant branch. As we shall show below, the origin of a red giant branch is closely related to the existence of the Schönberg-Chandrasekhar limit.

After hydrogen core exhaustion, a star develops an isothermal core that is surrounded by a hydrogen-burning shell. It can be shown by means of numerical integration of the equations of stellar interiors that if the mass of a nondegenerate hydrogen exhausted core exceeds ~ 12 percent of the total stellar mass, an equilibrium model with a hydrogen-burning shell source and nondegenerate isothermal core cannot be constructed. This result can be qualitatively explained by means of the virial theorem [see Equation (3-19)], which requires

$$4\pi P_1 r_1^3 = 3(\gamma - 1)U_1 + \Omega_1 \tag{5-7}$$

where

$$U_1 \sim \frac{3}{2} \frac{kT_c M_{core}}{m_p \mu}$$

$$\Omega_1 \sim -\frac{GM_{core}^2}{r_1}$$

and r_1 is the radius of the core. Equation (5–7) implies that the external pressure on a *nondegenerate* core of fixed mass and temperature has an extremum in r_1. We can estimate this extremum if we set

$$\left(\frac{dP}{dr}\right)_{r=r_1} = 0$$

and then use this condition to eliminate P_1 from Equation (5–7). We find

$$r_1 = \frac{4}{9} \frac{GM_{core}}{T_c} \frac{\mu m_P}{k}. \tag{5–8}$$

Since

$$\left(\frac{d^2P}{dr^2}\right)_{r=r_1} < 0$$

it follows that the isothermal core can support a maximum external pressure, which is

$$P_{max} \sim \frac{1}{G^3} \left(\frac{kT_c}{m_P\mu}\right)^4 \frac{1}{M_{core}^2} \tag{5–9}$$

where μ is the molecular weight of the core, and T_c is the temperature of the core. The high-temperature sensitivity of the nuclear burning source (in this case the CNO cycle) implies that the value of T_c does not vary much. Equation (5–9) shows that as the mass of the core is increased, the maximum allowable pressure at the boundary of the core decreases. Detailed numerical calculations show that the pressure at the boundary of the core exceeds the maximum pressure for equilibrium when $M_{core} \gtrsim 0.12$ M where M is the total mass of the star. These same calculations show that if $M \lesssim 5$ M$\odot$, the core can become $\sim$ isothermal after hydrogen exhaustion. On the other hand, if $M \gtrsim 5$ M$\odot$, the core will not become isothermal after hydrogen core exhaustion, but will contract immediately.

Large Radii and Luminosities

After hydrogen core exhaustion, a star of moderate mass will consist of a nearly isothermal core and a thin hydrogen-burning shell. Since the pressure gradient must support the weight of the core, the absence of a temperature gradient means that the density gradient inside the isothermal core must be correspondingly large. As we shall show below, the radius of an isothermal core remains almost constant as its mass is increased. This result implies that the density inside the core must increase as the shell advances. To show that this is the case, we write

$$\frac{dT}{dr} = \frac{dT}{dP}\frac{dP}{dr} = \frac{-T\rho GM_r}{P(n+1)r^2} \tag{5–10}$$

where

$$n + 1 \equiv \frac{d \ln P}{d \ln T}$$

and

$$P = \frac{\rho k T}{\mu m_P}.$$

If we assume that M_r and $n + 1$ are constant and integrate Equation (5–10) from the hydrogen-burning shell to larger r, we find

$$T - T_1 \sim \frac{\mu G M_r m_P}{(n + 1)k} \left(\frac{1}{r} - \frac{1}{r_1} \right). \tag{5–11}$$

The relation (5–11) implies

$$T_1 \sim \frac{\mu G M_r m_P}{(n + 1)k r_1} \tag{5–12}$$

if $r \gg r_1$, and $T \ll T_1$. The relation (5–12) shows that because T_1 is nearly constant, r_1 will remain nearly constant as the hydrogen shell advances. This result is borne out by detailed calculations.

The very large density gradient that is necessary to support an isothermal core implies that

$$\frac{d \ln M_r}{d \ln r} = \frac{4 \pi r^3 \rho}{M_r} \tag{5–13}$$

will be small outside the core. Integrating Equation (5–13), we find

$$\ln R = \int_{M_{\text{Shell}}}^{M} \frac{d \ln r}{d \ln M_r} d \ln M_r + \ln r_1. \tag{5–14}$$

Equation (5–14) shows that since r_1, the radius of the hydrogen-burning shell, is approximately constant ($\sim 2 \times 10^9$ cm), the reduction in $(d \ln M_r / d \ln r)$ that follows an increase in the mass of the isothermal core can lead to greatly increased total radius ($R \sim 10^{13}$ cm). Detailed numerical calculations show that this is indeed the case.

If we assume that $n + 1$ in Equation (5–10) is approximately constant outside the hydrogen-burning shell, we can derive an approximate expression for the luminosity of a thin shell. We can write

$$n + 1 = \frac{T}{P} \frac{dP}{dr} \frac{dr}{dT}. \tag{5–15}$$

If we use Equations (2–37) and (2–43) to substitute for dP/dr and dT/dr in Equation (5–15) and make the further substitution $\frac{1}{3} a T^4 = (1 - \beta)P$, we find

$$L_r = \frac{G M_r 16 \pi c (1 - \beta)}{\kappa (n + 1)} \tag{5–16}$$

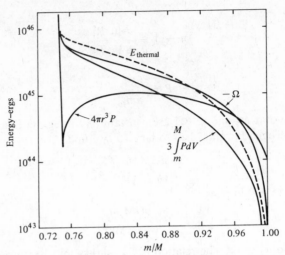

Figure 5-2   Various virial theorem terms are shown for a red giant envelope. $\Omega(m)$ is the gravitational energy between the m mass shell and the surface M. E_{thermal} is the corresponding amount of thermal energy (including energy of ionization).

Source: From Rose, W. and R. Smith. 1972, *Ap. J.*, **173**, 389, by permission of the University of Chicago Press.

where $n \sim 3$ and $\kappa \sim 0.4$ in the radiative layer that includes the hydrogen-burning shell. The virial theorem terms for a luminous red giant envelope are shown schematically in Figure 5-2.

5-4   THERMAL INSTABILITY

The question of the stability of nuclear burning sources within a star is a very old one. The observed secular stability of the sun on geological time scales indicates that it behaves as a self-regulated system. Any increase (decrease) in the rate of nuclear energy generation will be compensated for by an expansion (contraction) that will return the overall rate of nuclear energy generation to its unperturbed value. Jeans argued that main-sequence stars such as the sun should be thermally stable (that is, the rate of nuclear energy generation should be essentially constant over a Kelvin time scale). However, theoretical studies have shown that during some evolutionary stages, nuclear energy generation can take place in a thermally unstable manner and thereby produce significant evolutionary changes in a Kelvin time scale.

The ignition of nuclear burning under degenerate conditions was the first suggested example of thermal instability in stars. Such burning is expected to lead to very high rates of nuclear energy generation because the pressure and density of degenerate matter are relatively insensitive to changes in temperature, and in addition the nuclear burning rates of the relevant nuclear reactions are highly temperature sensitive (for example, $\varepsilon_{3\alpha} \propto T^n$, $n \sim 30$; and $\varepsilon_{\text{CNO}} \propto T^n$, $n \sim 15$). If the ignition of nuclear burning arises under degenerate conditions, expansive cooling may not accompany increased rates of nuclear energy generation until degeneracy is lifted or an explosion occurs. The onset of helium core burning will take place under degenerate conditions in stars less massive than about 2 M$\odot$. More recently it has been shown that if a direct $e^- - \nu$ interaction exists, carbon core burning is likely to arise under explosive conditions in stars of intermediate mass (that is, 1.4 M$\odot \leq M \leq$ 9 M$\odot$).

Recent studies of models of helium and hydrogen shell burning stars have shown that degeneracy is not a necessary condition for unstable burning of nuclear fuel in stars. The calculations have shown that during the second ascent of a star in the red giant branch, the burning of helium in a shell takes place under thermally unstable conditions, which implies that the rate of helium burning increases on a local Kelvin time scale until some limiting amplitude is attained. Figure 5–3 shows examples of relaxation oscillations produced as a result of thermal instability.

Hydrostatic equilibrium must be maintained during the development of a thermal instability because the condition $\tau_D \ll \tau_K$ is satisfied. For a star that is nearly chemically homogeneous, the requirement of hydrostatic equilibrium implies

$$\frac{\delta P}{P} = \frac{4}{3} \frac{\delta \rho}{\rho} \qquad (5\text{–}17)$$

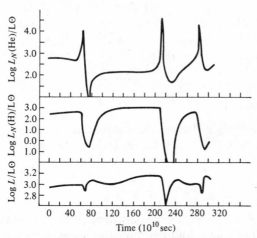

Figure 5–3   The helium burning luminosity $L_N(\text{He})/\text{L}\odot$, the hydrogen burning luminosity $L_N(\text{H})/\text{L}\odot$ and the surface luminosity $L/\text{L}\odot$ are shown as a function of time for a red giant whose helium burning shell is thermally unstable.

which reduces to

$$\frac{\delta T}{T} = \frac{1}{3}\frac{\delta \rho}{\rho} \qquad (5\text{--}18)$$

if the equation of state is that of a perfect nondegenerate gas. Equation (5–18) shows that a homogeneous star must cool as it expands, and consequently if the rate of nuclear energy generation is increased beyond its equilibrium value, the expansion will produce cooling and thereby decrease the rate of nuclear energy generation.

Although nondegenerate stars are stable against perturbations such as those given in Equations (5–17) and (5–18), they may become thermally unstable if perturbations that preserve hydrostatic equilibrium produce only a small fractional change in the pressure inside a nuclear burning shell source. If the fractional change in pressure is small, then

$$\frac{\delta \rho}{\rho} \sim - \frac{\delta T}{T}. \qquad (5\text{--}19)$$

For this type of perturbation, a temperature rise can accompany expansion, and consequently thermal instability can arise.

If the nuclear energy source is the triple-α reactions or the CNO cycle, then the rate of nuclear energy generation is highly temperature sensitive, and a thermal runaway can be initiated in an expanding shell source so long as the energy input into the burning region exceeds the radiation losses from the region, and the fractional changes in pressure remain small.

Let us assume that a burning shell source is sufficiently thin such that fractional perturbations in its density will produce much smaller fractional perturbations in its pressure, and so perturbations arise under isobaric conditions. The equation of radiative transfer can be written:

$$\frac{L_r}{4\pi r^2} = - \frac{4acT^3}{3\kappa\rho}\frac{dT}{dr} = -\sigma\frac{dT}{dr} \qquad (5\text{--}20)$$

where $dM_r = 4\pi r^2\rho\,dr$. Since the temperature sensitivity of the triple-α reactions and the CNO cycle are quite high, we can write

$$\frac{\delta\varepsilon}{\varepsilon} \simeq \nu\frac{\delta T}{T}. \qquad (5\text{--}21)$$

Assume that the initial temperature distribution in the shell has peak value T at center of shell source and that radiative losses occur equally from both sides which have fixed temperature T_0. If we further assume that the variation in the temperature gradient is much greater than either the variation in σ

or r, then we obtain

$$\delta \frac{dL}{dM_r} \simeq \frac{2L_N}{\frac{1}{2}\Delta M} \frac{\delta T}{(T - T_0)} \tag{5-22}$$

where ΔM is the mass inside the shell, and $\varepsilon = L_N/\Delta M$. Equations (2–76), (5–21), and (5–22) imply

$$\left[v - \frac{4T}{(T - T_0)} \right] \frac{\delta T}{T} = \left[\frac{\frac{3}{2}(P/\rho)}{\varepsilon} \right] \frac{d \, \delta \ln (P/\rho^{5/3})}{dt} \tag{5-23}$$

where it is assumed that the unperturbed value of the time derivative given in Equation (2–76) is small. If the left-hand side of Equation (5–23) is positive, then the nuclear energy input will exceed the radiative losses, and a positive temperature perturbation will lead to an increase in the entropy. Equation (5–23) indicates that the e-folding time for the development of a thermal instability (that is, local Kelvin time) is

$$\tau \simeq \frac{\frac{3}{2}(P/\rho)}{\varepsilon}. \tag{5-24}$$

If we neglect changes in r and σ as well as the time derivative in Equation (2–76), Equations (2–76) and (5–20) can be combined to give

$$\sigma \frac{d^2 T}{dr^2} = -\rho\varepsilon. \tag{5-25}$$

For the triple-α reactions,

$$\varepsilon_{3\alpha} = 3.9 \times 10^{11} f \rho^2 Y^3 \left(\frac{10^8}{T} \right)^3 e^{-A/T}.$$

f, the screening factor, is usually of order unity; Y is the helium mass fraction; and $A = 42.9 \times 10^8$. If $(T - T_0) \ll T$, and $T_0 \ll 1.5 \times 10^9$, then

$$\varepsilon = \varepsilon_0 e^{A(T - T_0)/T_0^2}. \tag{5-26}$$

The theory of thermal explosions asserts that the absence of stationary solutions to Equation (5–25) determines the necessary condition for a thermal explosion. If l is the thickness of the helium-burning shell (see Figure 5–4), it can be shown that this requirement leads to the condition

$$\lambda \equiv \left(\frac{\rho\varepsilon_0}{\sigma} \right) \left(\frac{A}{T_0^2} \right) \left(\frac{l}{4} \right)^2 \gtrsim 1 \tag{5-27}$$

for instability to occur. If the computed physical parameters for a helium shell burning star are substituted into the inequality (5–27), the condition for instability is satisfied by a factor of order unity. An equivalent criterion for instability can be obtained by comparing the diffusion time τ_{diff} for energy to leave the shell with the thermal e-folding time τ. Instability is predicted if $\tau < \tau_{\text{diff}}$.

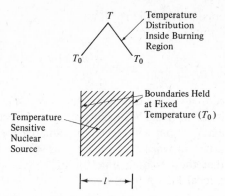

Figure 5–4   The temperature distribution inside a burning region whose boundaries are held at a fixed temperature T_0.

s-Process Nucleosynthesis

Because many isotopes are not likely to be formed during the principal nuclear burning phases, it is clear that several physical mechanisms are required in order to explain the formation of heavy elements. In the s process, the buildup of heavy elements is caused by the capture of neutrons on a time scale that is long as compared to the relevant beta decay lifetimes. This mode of nucleosynthesis is believed to be responsible for the production of many isotopes with atomic number A in the range 56 to 209.

An isotope of charge Z and atomic number A exposed to a neutron flux will undergo the reaction

$$(Z, A) + n \rightarrow (Z, A + 1) + \gamma. \tag{5–28}$$

If the isotope on the right-hand side of the equation is stable against beta decay, it will remain in this state until the next neutron capture. On the other hand, if it is radioactively unstable, it may beta-decay to $(Z + 1, A + 1)$ or capture a second neutron depending on the relative time scales for beta decay and neutron capture. s-process nucleosynthesis arises whenever the time scale for the next neutron capture is much longer than the relevant beta decay lifetimes.

It is likely that iron peak nuclei especially Fe^{56} are the seed nuclei for s-process nucleosynthesis. The equation that governs the buildup of the abundances N_A of an isotope of atomic number A as function of time is

$$\frac{dN_A}{dt} = n(t)\bar{v}[-\sigma_A(T)N_A + \sigma_{A-1}(T)N_{A-1}] \tag{5–29}$$

where $n(t)$ is the number density of neutrons, $\bar{v}$ their mean velocity, and σ_A and σ_{A-1} the neutron capture cross sections. The relatively small neutron cross sections that exist in the vicinity of closed nuclear shells (magic numbers) is the most important factor in determining s-process nucleosynthesis. Since Equation (5–29) predicts that N_A will decrease if the first term inside the brackets dominates the second term and conversely, solutions to Equation (5–29) will tend to minimize the right-hand side of the equation and thereby lead to the relation

$$\sigma_A N_A \simeq \sigma_{A-1} N_{A-1} \tag{5–30}$$

unless the isotope is close to a closed shell where σ is small. Observed abundances and measured nuclear cross sections for a number of isotopes such as Sm^{148} and Sm^{150} indicate that relation (5–30) is correct.

Several elements (for example, Zr, Ba, and technetium) that are predicted to be produced by means of the s process are observed to be overabundant in some red giants. It is, therefore, of great interest to find suitable sites for s-process nucleosynthesis in red giants. During a thermal relaxation oscillation, a convective zone extends from the helium-burning shell into the helium-rich layer between the helium- and hydrogen-burning shells. This convective zone may extend into the hydrogen-rich layer above and thereby mix a small amount of hydrogen into the helium-rich zone. Since the helium-rich layer contains significant amounts of C^{12}, C^{13} will be produced by means of the reaction,

$$p + C^{12} \rightarrow C^{13} + e^+ + \nu.$$

The isotope C^{13} will then undergo the reaction:

$$C^{13} + \alpha \rightarrow O^{16} + n \tag{5–31}$$

and produce a neutron that can interact with iron peak elements and produce s-process isotopes. It is plausible that a succession of neutron fluxes such as is required to explain the observed abundances of s-process isotopes could be produced by means of thermal relaxation oscillations.

5–5 LUMINOUS RED GIANTS

If stars did not lose significant mass, most would become either neutron stars or collapsed objects called black holes rather than white dwarfs. Observational evidence indicates that most mass loss occurs from luminous red giants. It is, therefore, of great interest to look for a physical basis for the occurrence of mass loss in these stars.

As a star evolves up the red giant branch, it becomes more luminous and its convective envelope more extended. The extended envelopes of red giants are divided into an outer region that is in radiative equilibrium, a region in which convective energy transport is fairly efficient (this region, which contains most of the mass, extends from the hydrogen ionization zone inward to a temperature of about $10^{5°}$K), and an inner region in which convection is inefficient and therefore high radiative fluxes are likely to cause departures from hydrostatic equilibrium.

Since they are extremely tenuous, the envelopes of luminous red giants contain temperature gradients which are significantly superadiabatic. From the definition of the local polytropic index n, that is,

$$1 + \frac{1}{n} \equiv \frac{d \ln P}{d \ln \rho} \tag{5-32}$$

and the perfect gas law

$$P = \frac{\rho k T}{\mu m_p} + \frac{1}{3} a T^4 \tag{5-33}$$

we obtain

$$n + 1 = \left(\frac{d \ln T}{d \ln P} - \frac{d \ln \beta}{d \ln P} - \frac{d \ln \mu}{d \ln P} \right)^{-1} \tag{5-34}$$

where β is the gas pressure divided by the total pressure. Equation (5-34) shows that large superadiabatic gradients (that is, $d \ln T/d \ln P$ large) produce small values of n and therefore relatively flat density distributions. The pulsation modes of such envelopes will have large amplitudes well below the surface.

The dynamical time scales of very luminous red giants are nearly as long as their characteristic thermal time scales. This circumstance suggests that deviations from hydrostatic equilibrium are likely, since a necessary condition for hydrostatic equilibrium is that the inequality $\tau_K > \tau_D$ be satisfied. Moreover, the theory of pulsational instability shows that the pulsational growth rate is about equal to a Kelvin time scale [see Equations (4-13) and (4-14)], and consequently the pulsational growth rates of luminous red giants are likely to be significant in one dynamical time scale.

Hydrodynamic calculations indicate that relaxation oscillations are excited in the envelopes of luminous red giants and that these oscillations, which are very strong pulsations, cause significant mass loss. Two physical mechanisms for driving such oscillations are predicted on the basis of stellar evolutionary calculations: (1) high radiation pressure gradients at the base of the envelope that are caused by thermal instability in the core and (2) instability to nonadiabatic pulsations.

The mass loss mechanism works as follows. The unstable red giant envelope expands initially. This expansion, which may cause a fraction of the mass of the envelope to be lost, is reversed by radiative losses, and the envelope collapses. A shock wave that is formed as a result of collapse propagates outward and ejects a fraction of the mass of the envelope. Numerical calculations indicate that the entire envelope can be ejected in a relatively short (10^2–10^3 yr) time scale. A star whose initial mass is about one solar mass will thereby end up as a white dwarf of about 0.7 M$\odot$.

The excitation of *envelope* relaxation oscillations is caused by the diffusion of radiation from the helium-burning shell to the convective envelope after a period of very high nuclear energy generation. The luminosities produced at the base of the convective envelope are too high for hydrostatic equilibrium to be maintained. In the interior of a star, very-high-energy fluxes are generally transported by means of convection rather than radiative transport. However, at the base of the envelope, convection cannot be efficient, since the maximum convective luminosity

$$(L_c)_{\text{max}} \simeq 4\pi r^2 \rho E v_s \tag{5-35}$$

(where v_s is the sound velocity and E is the internal energy per unit gram) is considerably less than the total luminosity of a luminous red giant.

The equation of hydrostatic equilibrium (2–36) and radiative transport (2–43) can be combined to give

$$\frac{P_g}{\rho}\frac{d\rho}{dr} + \frac{P_g}{\mu}\frac{d\mu}{dr} = \frac{1}{r^2}\left[\frac{\kappa\rho L_R}{4\pi c}\left(\frac{P_g}{4P_R} + 1\right) - GM_r\rho\right] \tag{5-36}$$

where P is the gas pressure; P_R is the radiation pressure; L_R is the radiative luminosity; and μ is the molecular weight. Since the molecular weight is constant at the *base* of the envelope where ionization is complete, Equation (5–36) can be rewritten as

$$\frac{d\rho}{dr} = \frac{1}{r^2}\frac{\mu m_p \rho}{kT}\left[\frac{\kappa L_R}{4\pi c}\left(\frac{P_g}{4P_R} + 1\right) - GM_r\right]. \tag{5-37}$$

The derivative $d\rho/dr$ is normally negative. Equation (5–37) shows that for L_R sufficiently large, $d\rho/dr$ would have to become positive for hydrostatic equilibrium to be maintained. This condition cannot be satisfied throughout the base of envelope since the density is higher in the core and hydrostatic solutions cannot be obtained. For this reason, hydrodynamic expansion must take place.

In Sections (5–4) and (5–5) we emphasized the role that global instabilities are likely to play in causing mass loss from red giants. However, mass loss is also likely to be a consequence of local turbulence or radiation pressure at the surface (see Section 2–9). Regardless of the relative importance of such local mechanisms, stellar model calculations have provided sufficient cause for most stars to lose a large fraction of their mass, and consequently it is possible to understand why most stars must become white dwarfs rather than neutron stars or black holes.

CHAPTER 6
Planetary Nebulae

6–1 GENERAL DESCRIPTION

Planetary nebulae are clouds of ionized gas (diameter $\sim 10^{17}$ cm) that surround and appear to be expanding from hot central stars. Some appear as pale green disks that resemble Neptune and Uranus. Most of the optical radiation that we observe from planetary nebulae was initially emitted by the hot central star ($T_s \sim 10^{5\circ}$K) in the ultraviolet (UV), absorbed by the highly ionized gas ($T \sim 10^{4\circ}$K), and then reradiated as visual radiation. For a typical planetary nebula, the stellar radiation is diluted by a factor $W = R_s^2/4R^2 \sim 10^{-14}$ where R_s is the radius of the star and R the radius of the nebula. The electron densities are $\sim 10^4$ cm^{-3}, and the helium abundance is estimated to be ~ 30 percent by mass. Diffuse nebulae such as Orion are similar to planetary nebulae in many respects. However, the distinguishing feature of planetary nebulae is the close connection between the nebula and the central star.

Observational studies of planetary nebulae aim to interpret the spectrum in terms of physical processes taking place in a low-density gas exposed to UV radiation and to obtain information about the temperature, density, chemical composition, and dynamical state of this gas. In addition, it is of great interest to determine the physical characteristics of the central star. The measurable quantities associated with planetary nebulae include angular size, surface brightness, the relative strengths of emission lines, the strength and spectrum of the radio emission, estimates for the luminosity, and effective temperatures of the central stars.

The total number of planetary nebulae within the galaxy is about 10^5. One is known to be within a globular cluster (M15). The distribution of planetary nebulae with respect to galactic coordinates indicates that they belong both to population I and population II. Unfortunately, the distances to planetary nebulae can only be estimated approximately, since their parallaxes are too small, and proper motion data combined with radial velocity data can only give a crude indication of distance. For example, the radius of the ionized region might expand at a different rate than the physical radius of the shell.

The spectra of planetary nebulae show that they are expanding. The spectral lines are split near the center of the nebulae and fused toward the edges. These observations indicate that, in the center of the nebula, the front side is moving toward us, and its lines are displaced toward the violet. On the other hand, the back side of the nebula is receding, and consequently its lines are displaced toward the red. The inferred expansion velocities are ~ 20 km/sec. What we observe is apparently either an expanding shell or ring. It is of interest to note that the observed expansion velocities for planetary nebulae are comparable to the escape velocities from the surfaces of red giant stars. There are differences in the expansion velocities for weakly ionized atoms (for example, O I, O II, S II, N II) and more highly ionized atoms (Ne III, O III). Because of strong frictional forces, different ions would move with the same velocities if they were in the same region. This means that different ions must lie in different regions of nebulae and that these different regions must be expanding at different velocities. Multi-ionized ions, in general, radiate from regions closer to the central star than singly ionized atoms, which means that the velocity of the outer regions of the nebula are expanding more rapidly than the inner regions. The presence of sharp outer boundaries in young planetary nebulae provides evidence that the mass ejection was fairly rapid ($\lesssim 10^3$ yr).

The components of the split lines in the central part of the nebula are narrow (that is, they have a width that is less than the splitting). This implies that the velocity dispersion inside the nebulae is considerably less than the speed of expansion so that the motion is essentially laminar flow without excessive turbulent motion.

6-2 PHYSICAL PROCESSES IN PLANETARY NEBULAE

Photoionization and Recombination

Far ultraviolet radiation from the central star photoionizes atoms and ions in the nebula. For example,

$$H + h\nu \rightarrow p + e^- \qquad (h\nu > 13.6 \text{ eV}).$$

The excess kinetic energy of the ejected electron is shared with other electrons by means of e^--e^- scattering before recombination can take place. This means that photoionization can heat the gas.

The cross section for photoionization from the energy level with principal quantum number n of a hydrogenic atom with nuclear charge Z can be written

$$\sigma_n(Z, \varepsilon_e) = \frac{2^6 \pi \alpha a_0^2}{3\sqrt{3}} \frac{n}{Z^2} (1 + n^2 \varepsilon_e)^{-3} g(n, \varepsilon_e) \qquad (6\text{-}1)$$

where a_0 is the Bohr radius; $\alpha = \frac{1}{137}$; g is the Gaunt factor which is of order unity; and $\varepsilon_e Z^2$ is the excess kinetic energy of the electron in units of the hydrogen ionization potential. The frequency of the ionizing photon and the excess kinetic energy of the electron are related by the expression

$$hv = Z^2 R_y \left(\frac{1}{n^2} + \varepsilon_e \right) \tag{6-2}$$

where $R_y = 13.6$ eV is the hydrogen ionization potential (that is, Rydberg).

Photoionized electrons will eventually recombine and thereby produce recombination lines,

$$p + e^- \rightarrow H(n, l) + hv.$$

The principle of detailed balance can be used to relate the recombination rate, which is in units of cm^{-3}-sec^{-1}, to the photoionization cross section given above. In thermodynamic equilibrium, the total rate of recombination to level n must equal the rate of photoionization from level n. This condition implies

$$n_p n_e \alpha_n(T) = n_H \int_{R_y}^{\infty} U_v c \sigma_n(v)(1 - e^{-hv/kT})\, dv \tag{6-3}$$

where the factor $(1 - e^{-hv/kT})$ on the right-hand side of the equation is due to induced emission, and U_v is the density of blackbody radiation per unit frequency interval (see Appendix A). From Equations (6–3) and the Saha equation [Equation (2–10)] it follows that

$$\alpha_n(T) = \left(\frac{2}{\pi} \right)^{1/2} \frac{g_n \exp(R_y/kT)}{c^2 (mkT)^{3/2}} \int_{R_y}^{\infty} (hv)^2 \exp\left(-\frac{hv}{kT} \right) \sigma_n(v)\, d(hv) \tag{6-4}$$

where g_n is the degeneracy of level n.

Collisional Excitation

Electrons can excite ions from their ground state to low-lying metastable levels. If the density of the gas is sufficiently low, de-excitation of these levels can take place either by means of magnetic dipole or electric quadrupole radiation rather than collisional de-excitation. Few permitted levels can be collisionally excited in planetary nebulae because their energies are too far above the ground state.

The strongest observed low-lying metastable transitions are those of [O III]. These transitions, which were originally attributed to an unknown element called nebulium before the correct spectral identifications were made, are responsible for the pale green color of planetary nebulae. Figure

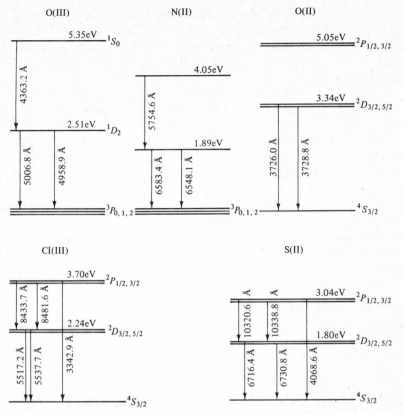

Figure 6-1   Energy level diagram of ions most important in the production of forbidden lines in planetary nebulae.

6-1 gives the energy level diagram for the lowest levels of nebulium and several other important ions. Collisional excitation of the ^1D state

$$O^{+2}\ (2p^2\ ^3P) + e^- \rightarrow O^{+2}\ (2p^2\ ^1D) + e^-$$

is followed by the two nebulium transitions

$$N_1\ 5007\ \text{Å}\ ^1D \rightarrow {}^3P_2$$
$$N_2\ 4959\ \text{Å}\ ^1D \rightarrow {}^3P_0.$$

Other forbidden transitions such as [O I], [O II], [N II], and [S II] are commonly observed in planetary nebulae. Since the relative intensities of emission lines depends on the temperature and electron density inside the nebula, estimates for these quantities may be obtained by studying the ratios of the intensities of a number of emission lines.

If an excited state is to be populated by means of collisions with electrons, the kinetic energy of the electron which is $\sim kT$ must equal or exceed the energy difference between the ground state and the excited state in question. For this reason, the ratio of intensities of forbidden transitions will, in general, depend quite strongly on the temperature of the gas. The strong density dependence of the ratio of intensities from two excited levels can be understood if we compare the limiting cases of low and high density. If the density of the gas is sufficiently low, all collisionally induced excitations will lead to radiative decay, and consequently the ratio of intensities will depend only on the rate of collisional excitation. On the other hand, if the density is sufficiently high, collisional excitation and de-excitation will be sufficiently rapid for the population ratios to satisfy the usual Boltzmann distribution; consequently the intensity of emission lines will be proportional to the Boltzmann factor times the radiative transition probability.

Measurements of emission lines are also of considerable importance in determining the physical conditions inside nova shells, supernova remnants, quasistellar objects, and peculiar galaxies.

Radio and Infrared Emission

The absorption coefficient for free-free absorption can be written

$$k_v = \frac{A(T, v)n_e^2}{v^2 T^{3/2}} \qquad (hv \ll kT) \qquad (6\text{--}5)$$

where $A(T, v)$ varies slowly with T, and Equation (6–5) follows from Equation (1–21) and the principle of detailed balance. The measured radio flux density due to free-free emission from an ionized gas is

$$F_v = \frac{2kT_b v^2}{c^2} \Delta\Omega \qquad (6\text{--}6)$$

where T_b is the brightness temperature of the nebula, and $\Delta\Omega$ is its solid angle. If the planetary nebula is sufficiently transparent to radio radiation, we have

$$T_b \simeq T\tau_v = T \int_0^L k_v \, dl \qquad (6\text{--}7)$$

where L is the linear dimension of the nebula. It is important to emphasize that for thermal emission, the brightness temperature cannot exceed the kinetic temperature of the gas. The optical depth becomes

$$\tau_v = \frac{A(v, T) \int_0^L n_e^2 \, dl}{v^2 T^{3/2}} \qquad (6\text{--}8)$$

where the integral in Equation (6–8) is called the emission measure. Equations (6–6), (6–7), and (6–8) allow us to express the emission measure in terms of quantities that can be either measured or calculated. We find

$$E \equiv \int_0^L n_e^2 \, dl = \frac{F_v T^{1/2} c^2}{2kA(T, v) \, \Delta\Omega}.$$ (6–9)

Equation (6–9) shows that the mean densities and masses of planetary nebulae can be estimated if the distance and filling factor (that is, fraction of the volume of nebula occupied by mass) are known.

The ratio of the $H\beta$ flux to radio flux can be used to estimate the temperature of the gas if the radio emission is assumed to be due to free-free emission or to check the thermal nature of the radio emission. Nearly all planetary nebulae appear to be thermal sources.

Several planetary nebulae are observed to have large infrared excesses. The physical nature of this radiation is presently unknown. However, continuum radiation from dust is the likely physical process.

Spatial Distribution and Distances

The total luminosity of a planetary nebula in $H\beta$ radiation can be written

$$L(H\beta) = \frac{4\pi}{3} R^3 f n_e^2 \alpha(H\beta) \frac{hc}{\lambda}$$ (6–10)

where $\lambda = 4861$ Å; R is the radius of the nebula; f is some unknown filling factor; and $\alpha(H\beta)$ is the $H\beta$ recombination rate. The electron number density has been set equal to the number of ionized hydrogen atoms. If $F(H\beta)$ is the flux in units of ergs-cm^{-2}-sec^{-1} as measured at the telescope, we find

$$F(H\beta) = \frac{L(H\beta)}{4\pi r^2} 10^{-C_\lambda(r)}$$ (6–11)

where r is the distance to the nebula, and $C_\lambda(r)$ includes the effects of interstellar extinction. Equations (6–10) and (6–11) imply

$$r = \frac{24F(H\beta)}{[\theta^3 f \alpha(H\beta) n_e^2 (hc/\lambda) 10^{-C_\lambda(r)}]}$$ (6–12)

where $\theta = 2R/r$ is the measured angular size. The above equation shows that the distance to planetary nebulae can be estimated if the flux in $H\beta$, angular size, T, n_e, and filling factor can be determined. This method for estimating distances to planetary nebulae does not require that we make any assumptions about the optical depth of the nebula to Lyman continuum radiation. However, because of observational difficulties associated with determining the ratios of forbidden line intensities, this method can only be applied to planetary nebulae with high surface brightness.

If we assume that the nebula is fully ionized, the mass of the nebula is

$$M = \frac{4\pi R^3 f n_e m_p}{3} \tag{6-13}$$

where we have neglected helium. Equations (6–12) and (6–13) can be combined to give the following expression for the distance

$$r^5 = 24 \left(\frac{M}{4\pi m_p}\right)^2 \frac{\alpha(H\beta)hc}{\lambda\theta^3 10^{C_{\lambda}(r)} F(H\beta)f}. \tag{6-14}$$

The distance estimated in this manner is only sensitive to the $\frac{2}{5}$th power of the mass and therefore not very sensitive to uncertainties in this quantity. Angular size and optical flux data have been used to make a statistical determination of the distance to more than 500 planetary nebulae. Measured angular sizes and $H\beta$ fluxes can be used in conjunction with Equation (6–14) to estimate the distances. Typical masses (~ 0.4 M$\odot$) and filling factors (~ 0.7) estimated from observations of relatively close nebulae are assumed. The distance scale is then found from the requirement that the spatial distribution of planetary nebulae be consistent with that of stars throughout the galaxy. If we know the distances to planetary nebulae, then we can estimate the total number throughout the galaxy ($\sim 10^5$). The estimated number density and lifetime (20,000–40,000 yr) can then be used to predict the formation rate of planetary nebulae. This turns out to be comparable to the local white dwarf birth rate, which is $\sim 2 \times 10^{-3}$ kpc^{-3}-yr^{-1}.

6–3 TEMPERATURES OF CENTRAL STARS

In a dilute radiation field such as exists inside a planetary nebula, high-frequency quanta emitted by the central star are converted into quanta of lower frequency by the surrounding nebula. Since most of the radiation from the central stars of planetary nebulae is emitted in the far UV, it is impossible to determine the temperatures of the central stars directly. However, the number of Balmer photons emitted by the nebula will, in general, be $\leq$ the number of quanta emitted by the central star beyond the lower limit of the Lyman continuum. This circumstance allows us to find a lower limit for the temperature of the central star. The number of photons emitted beyond the Lyman continuum is

$$N_{Ly} = \frac{8\pi^2 R_s^2}{c^2} \int_{v_0}^{\infty} \frac{v^2\,dv}{[\exp{(hv/kT_s)} - 1]}. \tag{6-15}$$

From the observations of planetary nebulae, one can determine the ratios

$$A_i = \frac{L_i}{[v_i(\partial L/\partial v)_i]} \tag{6-16}$$

where L_i is the energy per unit time emitted by the nebula in the ith Balmer line and $(\partial L/\partial v)_i$ is the power per unit bandwidth emitted by the central star in the vicinity of the ith Balmer line. The quantity $(\partial L/\partial v)_i$ is given by

$$\left(\frac{\partial L}{\partial v}\right)_i = \frac{8\pi^2 R_s^2 h v_i^3}{c^2} \frac{1}{[\exp(hv_i/kT_s) - 1]}. \tag{6-17}$$

From the above equations, it follows that the condition that the number of Balmer quanta emitted by the nebula be less than the number of Lyman continuum photons is

$$\sum_i \frac{v_i^3 A_i}{[\exp(hv_i/kT_s) - 1]} \leq \int_{v_0}^{\infty} \frac{v^2\, dv}{[\exp(hv/kT_s) - 1]} \tag{6-18}$$

where v_0 is the low-frequency end of the Lyman continuum.

The summation on the left-hand side of the above inequality includes Balmer lines and Balmer continuum. The ratios A_i can be measured, and consequently the above inequality gives a lower limit to the temperature of the central star.

6-4 LUMINOUS RED GIANTS AND THE ORIGIN OF PLANETARY NEBULAE

During a star's first evolution up the red giant branch, its energy source is a hydrogen-burning shell, which surrounds an inactive helium core. This evolutionary stage is terminated by the ignition of helium burning in the core. Subsequently, the luminosity of the star decreases sharply, and it evolved through a phase of helium burning in the core and hydrogen burning in a shell. This evolutionary stage is associated with the origin of the horizontal branch. After helium core exhaustion, a helium-burning shell is formed, and the star ascends the red giant branch for the second time.

As the star evolves up the red giant branch, it enters a region of the H-R diagram occupied by long period variables. The structure of the envelopes of luminous red giants (see Figure 5-1) is very striking. It appears plausible (see Section 5-5) that the origin of planetary nebulae is associated with the dynamical properties of very luminous red giant envelopes as well as the presence of thermal instability.

Figure 6-2 shows some central stars of planetary nebulae plotted on the H-R diagram. It is likely that planetary nebulae are the ejected envelopes of luminous red giants, and their central stars the remnant cores of these red giants. The schematic path shown in Figure 6–2 denotes the typical path of a central star. This approximate path is deduced from the observations because the diameters of nebulae along the path are observed to increase on the average. If 20 km/sec is taken to be a typical nebula expansion velocity, then the evolutionary lifetimes are as shown in Figure 6–2.

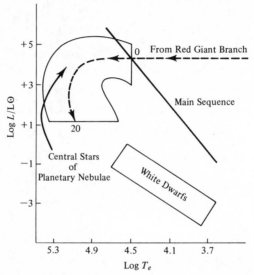

Figure 6–2 The schematic evolution of a star from the red giant branch into the white dwarf state is shown in the H–R diagram. It takes approximately 20,000 years for a star to evolve from the position denoted 0 to that denoted 20.

CHAPTER 7

Novae
and Galactic
X-Ray Sources

7-1 NOVAE

Observations

Novae are explosive variable stars that are observed to brighten suddenly by about 11 magnitudes and then decline. Before and after an outburst, novae are hot subdwarfs that are members of binary systems. At the observed maximum of an outburst, M_v is between -6 to -9, a value that is much less than that observed for a supernova outburst ($M_v \cong -18$). Although the initial rise to near maximum light is very rapid ($\lesssim 1$ day in some instances), the decline is much slower (years or, in extreme cases, decades). Most novae show variability during decline and even after they appear to have returned to their prenova state. There is good evidence that at the time of a nova outburst, a shell of matter is ejected from the star. This rapidly expanding shell produces the effect of an enlarged photosphere. Some novae are observed to recur, and it is generally assumed that most novae would repeat if they could be observed for a sufficiently long interval of time. The nova outburst is not a sharply defined phenomenon. It is clear that each nova outburst has its own peculiarities. For this reason, theories for novae must explain the broad outlines of the phenomenon and still be flexible enough to allow for the variations among novae.

Although the light curves of novae show great differences, similarities are apparent if they are appropriately normalized (see Figure 7–1). All show a rapid rise which is followed by a much slower decline. The initial rise of a nova to about 2 magnitudes below maximum is quite rapid. At this stage a premaximum halt which lasts for hours to days is often observed. The nova then brightens to maximum and declines. Some novae show strong variations in brightness during their decline while others fade more smoothly. These light fluctuations are generally more pronounced for fast novae as compared

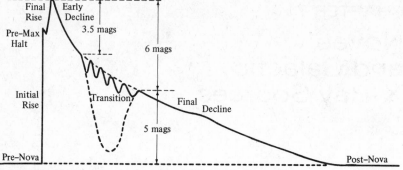

Figure 7-1   Various stages of a nova outburst are shown schematically as a function of time.

to those of moderate or slow rate of decline. The rate of decline from maximum is more rapid for novae with the brightest maxima, and consequently novae can be used as indicators of the distances to external galaxies. Recent measurements in the infrared and ultraviolet indicate that the decline from maximum is slower in these regions of the spectrum than in the visual region. These measurements imply that the total energy radiated during an outburst can be appreciably higher than the measured visual radiation which is typically 10^{44}–10^{45} ergs between maximum light and 3 magnitudes below maximum.

The development of the observed spectra of novae is related to the amount of decline from maximum light. At least five distinct types of spectra, which are associated with the various stages of the nova light curve, can usually be identified during an outburst. The expected line shape from an expanding shell of gas is an absorption feature displaced toward the blue and an emission peak broadened about the true wavelength.

The premaximum spectrum consists of moderately diffuse absorption without strong emission lines. The absorption spectrum usually becomes stronger and always changes to a later spectral class (A or F supergiant) as the nova rises to maximum light. The principal spectrum, which is characterized by strong, symmetrically broadened emission lines and shortward displaced absorption lines, appears immediately after maximum. The principal absorption spectrum has a greater displacement toward the blue than the premaximum spectrum. Emission lines that require relatively high excitation appear, while lines requiring lower excitation weaken during the development of the principal spectrum.

The diffuse enhanced spectrum, which is most prominent for slow novae, appears before the principal spectrum has lost much strength. Its displacement toward the blue is much greater than that of the principal spectrum. The diffuse spectrum shows an initially large spread in velocities that tends to sharpen and become multiple as a function of time. This circumstance provides evidence for continued ejection of mass after the main outburst. Mass ejected after the main outburst is believed to be responsible for the observed acceleration of the previously ejected mass shell.

The Orion spectrum appears next when the total luminosity is about 3 magnitudes below maximum. This spectrum is accompanied by a large increase in the color temperature and shows featureless emission and strongly displaced absorption lines that do not sharpen with time. Because large fluctuations in the intensity of the Orion spectral lines accompany fluctuations in the observed continuum radiation, it is likely that this spectrum arises from a region close to the star.

When the visual luminosity is about 4 magnitudes below maximum, nebula lines are observed. These forbidden transitions, which require relatively low density and high excitation temperatures, strengthen relative to permitted transitions. At about 7 magnitudes below maximum, the emission lines are similar to those of planetary nebulae except that their widths are much greater as a consequence of much higher velocities of expansion. The postnova star is usually variable. Sometimes this variability is periodic.

Novae occur with a frequency of about 50 per year per galaxy. Since the frequency of outburst is higher than the white dwarf formation rate, it is clear that most novae must recur. The spatial distribution of novae shows a marked concentration toward the galactic equator and center. The distances to novae can be estimated by measuring the Doppler shift and angular rate of expansion of the ejected mass shell. This determination of distance is made somewhat uncertain because the outburst is not generally spherically symmetric. Most novae do, however, show symmetry about an axis.

It is possible to estimate the amount of matter ejected during a nova outburst. To make this determination, one first estimates the electron temperature T of the mass shell by measuring the relative intensity of forbidden transitions (usually those of OIII). Since the emission of a recombination line such as Hα can be expressed in the form

$$E_{\mathrm{H}\alpha} = f(T)n_e{}^2 \cdot \text{volume} \qquad (7\text{-}1)$$

where n_e is the electron number density and $f(T)$ a known function of temperature, the mass of the shell ($\sim n_e m_p \cdot$ volume) can be estimated if the angular size, distance, and flux in Hα are measured. Estimates of the ejected mass are 10^{-4}–10^{-6} M$\odot$. The corresponding kinetic energies are 10^{43}–10^{45} ergs.

The connection between novae and binary membership is established. Most (probably all) novae are members of binary systems such that one component is a hot white dwarf and the companion a cooler star. The observations suggest that mass is flowing from the cooler to the hotter star. Such mass transfer provides a convenient mechanism for explaining the recurrence of novae.

Theory

Because nova outbursts are known to arise only in binary systems in which one component is a hot white dwarf, it has been suggested that mass accretion onto the surface of a white dwarf plays an essential role in the occurrence of novae. If a white dwarf accretes small amounts of hydrogen-rich mass, the envelope contracts. This contraction, which is nearly adiabatic, causes the temperature to increase until hydrogen burning becomes important. When the rate of nuclear energy generation is sufficiently high, the hydrogen-burning shell becomes thermally unstable (see Section 5–4). Mass accretion rates as low as 10^{-9} M$\odot$/yr are sufficient to produce such unstable hydrogen burning. Although the precise mass of the hydrogen-rich envelope necessary to initiate thermal instability will depend on the mass accretion rate as well as the mass of the white dwarf, it is likely to lie in the range of 10^{-4}–10^{-5} M$\odot$. The theory of white dwarfs provides a basis for understanding this result. More massive hydrogen-rich envelopes would imply that hydrogen exists at densities sufficiently high ($\geq 10^4$ g/cm^3) for hydrogen burning to take place even at zero temperature (see Section 8–4). On the other hand, less massive hydrogen-rich envelopes are unlikely to attain sufficiently high temperatures for hydrogen burning to become significant.

The luminosity produced by matter falling on the surface of a white dwarf is approximately

$$L = \frac{GM}{R}\frac{dM}{dt} \tag{7-2}$$

where dM/dt is the mass accretion rate, and GM/R, the gravitational energy per gram, is about 10^{17} ergs/g for a typical white dwarf. Equation (7–2) shows that even small amounts of mass accretion (10^{-8} M$\odot$/yr) can produce a luminosity that is greater than that of a typical white dwarf. Nova recurrence times are likely to be governed by the rate of mass accretion and the Kelvin time scale of the hydrogen-rich envelope. Since the ejected mass in a nova outburst is about 10^{-5} M$\odot$, mass accretion rates as low as 10^{-8}–10^{-9} M$\odot$/yr are sufficient to produce an outburst every 10^4–10^5 years.

The e-folding time for the nuclear burning to increase is given approximately by the local Kelvin time scale, that is,

$$\tau_K \sim \frac{3P/2\rho}{\varepsilon_{CNO}} \tag{7-3}$$

where the physical variables are evaluated at the mass point with the highest rate of nuclear energy generation per unit gram. Calculations show that the predicted high rates of nuclear energy generation cause the star to become pulsationally unstable. If the relative amplitude $\delta P/P$ of the fundamental pulsational mode is high at the position of the hydrogen-burning shell, it is relatively easy to excite pulsational instability by means of a temperature-sensitive nuclear energy source such as the CNO cycle. That this conclusion is correct follows from the stability integral given in Equation (4–13). Since the nuclear burning rate ε varies as about T^{15} for the CNO cycle, we find $\delta\varepsilon \sim 15\varepsilon(\delta T/T)$. If the stability integral is positive, pulsations will grow in amplitude. If it is negative, damping will take place. The nuclear term makes a positive contribution while the radiative loss term, which is more important near the surface, makes a negative contribution unless extensive ionization zones are present. The very short pulsation periods and the reasonably homologous character of the modes implies that it is possible to store a significant amount of energy in the radial mode of oscillation (10^{43}–10^{45} ergs).

It is not possible to determine the limiting amplitude by means of linear pulsation theory. Nonlinear hydrodynamic calculations have been carried out for pulsationally unstable models of Cepheids and RR Lyrae stars. Although the physical causes for the limit to the growth of pulsations in RR Lyrae stars are not fully understood, the calculations indicate that the leveling off of the excitation in the He^+ layer is the principal cause. For hot white dwarfs, the opacity is primarily electron scattering, and the contribution of ionization zones to the stability integral is negligible. Therefore, one would not expect nonlinear terms to significantly affect the excitation of pulsations until the amplitudes at the surface become sufficiently large for shock wave dissipation to become important. This theory suggests that a pre-nova star should pulsate before an outburst but not necessarily after an outburst.

In our discussion of pulsational instability, we considered only the radial modes of oscillation. However, the observational evidence indicates that most nova outbursts are not spherically symmetric. Moreover, many nova outbursts appear to have axial symmetry. Photographs of Nova Herculis illustrate this point. It is well known that a star has nonradial as well as radial modes of oscillation. The resonant excitation of nonradial modes by radial pulsations is one means by which these nonradial modes may become excited.

The character of the spectrum of nonradial modes is more complicated than the radial spectrum. It has been shown that in addition to a high-frequency branch of nonradial modes (p modes), a stable gas sphere will also have a branch whose periods increase indefinitely as the number of radial modes increases (g modes). The motion is primarily radial, and the pressure variations large for high-frequency oscillations. On the other hand, the motions are primarily horizontal for low-frequency oscillations. Modes of intermediate frequency (Kelvin modes) also exist.

It is likely that the rise to maximum of a nova is the result of the development of an extended photosphere that is subsequently ejected from the star. Assume that a spherical shell of matter is heated to such a high temperature that its gas pressure is negligible as compared to its radiation pressure. The luminosity of the expanding shell is

$$L(t) = -\frac{d}{dt} \int_0^{R(t)} aT^4(r, t)4\pi r^2 \, dr. \tag{7-4}$$

If an Eddington temperature distribution (see Appendix B) is assumed to hold throughout the shell, it follows that

$$T^4(r, t) = \frac{L(t)}{8\pi\sigma R^2} (1 + \tfrac{3}{2}\tau) \tag{7-5}$$

where the optical depth τ is

$$\tau = \frac{\kappa \, \Delta M(R - r)}{\tfrac{4}{3}\pi R^3}. \tag{7-6}$$

Equations (7-4) and (7-5) can be solved if the relation $R(t) = vt$ is assumed. The luminosity $L(t)$ is found to increase from some initial value to a maximum in a time

$$t_{max} = \left[\frac{3\kappa \, \Delta M}{16\pi vc} \frac{1}{1 + \tfrac{2}{3}(v/c)} \right]^{1/2} \tag{7-7}$$

and then decline monotonically in a time that is c/v times greater than the initial rise time. It follows from Equation (7-7) that the maximum luminosity occurs when the diffusion time scale $R\tau/c$ is approximately equal to the expansion time scale R/v. If plausible physical parameters for novae are assumed ($\kappa = 0.34$, $\Delta M = 10^{-5}$ M$\odot$, $v = 10^3$ km/sec), the predicted time to maximum luminosity is 4 h, which is comparable to that observed for the fastest known novae.

A satisfactory theory of novae should be able to explain their luminosities, which are used as a means of estimating the distances to external galaxies. Because the development of a thermal instability is limited by the expansion of the envelope, the total energy released must be less than the gravitational binding energy. This circumstance leads to the requirement that the nuclear energy release be less than

$$\frac{GM \, (M\text{-}M_{shell})}{R} \sim 10^{45}\text{-}10^{46} \text{ ergs} \tag{7-8}$$

where M is the total mass of the star and M_{shell} the mass interior to the hydrogen burning shell. The amount of energy radiated during a nova outburst is consistent with the above inequality. Likewise, the maximum luminosity

$$L_{max} \sim \frac{10^{45}}{t_{max}} \sim 10^6 \text{ L}\odot \tag{7-9}$$

is comparable to the maximum visual luminosities of the most luminous novae, which are observed to decline with the greatest rapidity.

An expanding gas cloud will eventually become optically thin. An optically thin gas cloud can radiate much more effectively than an optically thick cloud of comparable mean temperature. It is likely that after the initial rise of a nova outburst, particle ejection and radiation from the central star prevents the cloud from cooling rapidly. To show that an optically thin body can radiate more effectively than a comparable optically thick body, we recall that the effective temperature T_e need not be as high as the average temperature since

$$\sigma T_e^{\,4} \sim \frac{\lambda c E_r}{L} \sim \frac{\lambda}{L}\, \sigma T^4 \qquad (7\text{--}10)$$

where λ = mean free path of photon, L = characteristic length, E_r = the radiant energy density, and T = the average temperature. Equation (7–10) shows that if the photon mean free path is sufficiently short (that is, $(\lambda/L)^{1/4} \ll 1$), $T_e \ll T$. On the other hand, if $\lambda \sim L$, the effective temperature can approach the average temperature. The observations of novae suggest that maximum light occurs when the temperature in the outer layers falls below that necessary to ionize hydrogen, and consequently the opacity is greatly reduced.

Although the above discussion provides a basis for understanding some of the general characteristics of novae, the physical mechanism (or mechanisms) by which unstable nuclear burning ejects mass from a hot white dwarf has not been fully described. Hydrodynamic calculations indicate that there are at least two physical mechanisms by which mass loss can occur: (1) shock wave ejection and (2) pulsations.

If the rate of nuclear energy generation becomes sufficiently high, a strong shock wave (blast wave) will be formed. This shock wave will propagate outward and eject some fraction of the envelope. Such a strong shock wave will be formed if at some point in the envelope, the sound travel time to the surface exceeds the local Kelvin time and therefore the star is unable to adjust its structure to the rapidly changing pressure distribution. The ejected matter acquires a kinetic energy that is approximately

$$L_H \tau_D - |\Omega|$$

where L_H is the nuclear energy generation, τ_D the dynamical time scale, and $|\Omega|$ is the initial gravitational binding energy of the ejected envelope.

If the blast wave travels into a region of uniform density ρ_0, a self-similar solution that depends only on ρ_0 and the injected energy E can be found. This solution remains valid so long as the pressure P_1 behind the shock is much greater than the pressure P_0 in the undisturbed region ahead of it. The only dimensional combination of E and ρ_0 that contains only length and time is $E/\rho_0 = [\text{cm}^5\text{-sec}^{-2}]$. This implies that the dimensionless variable is

$$\xi = r\left(\frac{\rho_0}{Et^2}\right)^{1/5} \qquad (7\text{--}11)$$

and the motion of the wave front is governed by

$$R = \xi_0 \left(\frac{E}{\rho_0}\right)^{1/5} t^{2/5}. \qquad (7\text{--}12)$$

The propagation velocity of the shock wave becomes

$$D = \frac{dR}{dt} = \frac{2}{5}\frac{R}{t} = \xi_0 \frac{2}{5}\left(\frac{E}{\rho_0}\right)^{1/5} t^{-3/5}. \qquad (7\text{--}13)$$

For a strong shock, the values of the physical variables behind the shock front are

$$\rho_1 = \rho_0 \frac{\gamma + 1}{\gamma - 1} \qquad P_1 = \frac{2}{\gamma + 1}\rho_0 D^2 \qquad v_1 = \frac{2}{\gamma + 1} D \qquad (7\text{--}14)$$

where γ is the ratio of specific heats, c_p/c_v. Note that the density behind the front remains constant, but the pressure decreases as

$$P_1 \propto \rho_0 D^2 \sim \rho_0 \left(\frac{E}{\rho_0}\right)^{2/5} t^{-6/5} \sim \frac{E}{R^3}. \qquad (7\text{--}15)$$

The blast wave degenerates into a sound wave when $P_1 \to P_0$. The equations of continuity, motion, and adiabaticity enable the distributions of pressure, density, temperature, and velocity to be determined as a function of $\xi = \xi_0 r/R$. The unknown ξ_0 is found from the requirement that the total energy

$$E = \int_0^R 4\pi r^2 \left(\frac{P}{\gamma - 1} + \frac{\rho v^2}{2}\right) dr \qquad (7\text{--}16)$$

be constant.

The passage of a strong wave through an envelope takes place in a short time ($\simeq$ seconds). The hydrodynamical readjustment should take place in a time scale much shorter than that required for significant radiative losses. It would be difficult to explain the occurrence of post-maximum shells that can arise days or even weeks after an outburst if a strong shock wave is assumed to be the sole cause of the outburst.

Two beta decays (N^{13}, O^{15}) are necessary in order that the CNO cycle go to completion. The decay of N^{13} is the most important as far as determining the limiting rates of nuclear energy generation. Since beta-decay lifetimes are independent of temperature and density, it is clear that for a fixed amount of N^{13}, the burning rate cannot increase indefinitely as the temperature is increased. If solar abundances are assumed for the various isotopes that take part in the CNO cycle (see Section 2–7), the maximum rate of nuclear energy generation is about 10^{14} ergs-g^{-1}-sec^{-1} This burning rate is only marginally adequate to produce a strong shock wave in the envelope of a white dwarf. The presence of convective mixing between the shell source and the envelope implies that larger amounts of CNO isotopes are available to complete the CNO cycle and therefore higher rates of nuclear energy generation are plausible during the development of a thermal instability. The observed relatively high N^{14}/C^{12} and N^{14}/O^{16} abundance ratios observed in nova shells provide strong evidence that the CNO cycle is connected with the origin of novae since the CNO cycle tends to produce N^{14} at the expense of C^{12} and O^{16}.

Although the shock wave ejection mechanism is able to explain certain features of the nova outburst, it does not provide a completely satisfactory explanation for the velocity distributions of the ejected mass or the observed durations of outburst. Nuclear driven pulsations, which are expected to build up in amplitude until limited by shock wave dissipation near the surface, are an additional mechanism for mass ejection. This mechanism, which is likely to produce a stellar wind, is particularly promising for slow novae such as DQ Her, which is observed to pulsate. Pulsations can produce very extensive heating of the outer layers of a star. Such heating might initiate nuclear reactions and therefore provide an additional energy source for ejecting mass.

7–2 GALACTIC X-RAY SOURCES

The earth's atmosphere is highly opaque to x-ray radiation, and consequently observations of x-ray sources must be made from high-altitude balloons, rockets, or satellites. Such observations indicate that most discrete x-ray sources lie close to the plane of the galaxy and tend to be concentrated toward the galactic center. For this reason, it is likely that these objects are of galactic origin and also more distant than the characteristic thickness of the galactic plane (that is, > 200 pc). Highly condensed objects such as white dwarfs and neutron stars are the likely sources of this x-ray radiation. Black holes (see Section 11–4) are another interesting possibility.

Because it is very difficult to accurately determine the positions of x-ray sources and also because interstellar absorption is very great near the plane of the galaxy, prior to 1972 only two galactic x-ray sources had been securely identified with optically observed objects. The strongest known x-ray source called Scorpius X-1 is identified with a peculiar, variable blue star. X-ray emission is also observed from the Crab nebula. This emission has two components. One component, which shows no strong short-term variations, is extended in angular diameter. The other component comes directly from the neutron star near the center of the nebula.

Blackbody radiation, bremsstrahlung, and synchrotron radiation are the most likely x-ray emission mechanisms. In principle, observations of the spectra of x-ray sources make a discrimination between these mechanisms possible. Kinetic temperatures in excess of 10^{8}°K are required if the radiation is to be caused by thermal emission. White dwarfs and neutron stars are the only known type of star capable of producing such high temperatures. The predicted bremsstrahlung spectrum is flat up to some high frequency at which point there is an exponential cutoff (see Section 1–3). On the other hand, synchrotron radiation is expected to produce a power law spectrum (see Section 9–3). The observed spectrum of Scorpius X-1 is most consistent with thermal emission. X-ray emission from the Crab nebula is probably caused by synchrotron radiation (see Section 10–9). Time variations in flux are often observed from discrete x-ray sources. Although most time variations are quite random, some sources show periodic variability. The x-ray pulses from the Crab pulsar are characterized by a remarkably stable periodicity (0.033 sec).

Recent observations have shown that a number of x-ray sources are members of close binary systems. Two of these binary sources, Hercules X-1 and Centaurus X-3 vary periodically with periods of 1.2 and 5 seconds, respectively. These measurements indicate that mass accretion onto the surface of a condensed object from a companion star is a key physical process in producing some x-ray sources. In this respect the origin of novae and x-ray sources is similar. A rotating neutron star with a strong dipolar magnetic field is a likely clock mechanism for periodically varying x-ray sources. A pulsating white dwarf is another possibility.

CHAPTER 8
White
Dwarfs

8-1 LUMINOSITIES, RADII, AND MASSES

As compared with normal stars, white dwarfs are observed to have very low luminosities. On the other hand, their effective temperatures are often relatively high ($T_e \sim 10^{4\circ}$K). Such configurations must have very small ($\lesssim 10^9$ cm) radii and correspondingly high interior densities since their masses are about 1 M$\odot$. The basic theoretical problem associated with the existence of white dwarfs is to explain how such configurations, which must have very strong gravitational fields, can maintain hydrostatic equilibrium but have relatively low luminosity. It is clear that a white dwarf cannot be supported by a nondegenerate gas since the high-pressure gradients necessary to balance gravity would imply very-high-temperature gradients and therefore by Equation (2–43) very high surface luminosities. The Pauli exclusion principle implies that a degenerate electron gas can have a large kinetic energy per electron and therefore a high pressure even at zero temperature. In white dwarfs the pressure gradients necessary to support the structures are produced primarily by the zero point kinetic energy of a degenerate electron gas.

If it is assumed that the white dwarf equation of state is that of a free electron degenerate gas, a well-defined mass-radius relation must exist for white dwarfs with $L \lesssim 1$ L$\odot$ (see Figure 8–1 and the discussion that follows). The existence of a mass-radius relation allows us to estimate the mass of a white dwarf if we can estimate its distance and thereby determine its luminosity. The distance to more than 25 white dwarfs has been reliably determined by means of trigonometric parallaxes. Galactic cluster membership and membership in binary systems that appear to have main-sequence companions make it possible to estimate the distance to an additional ~ 100 white dwarfs. These observations indicate that the masses of white dwarfs range from ~ 0.3 M$\odot$ to ~ 1.2 M$\odot$ with 0.7 M$\odot$ typical.

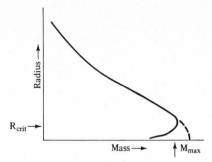

Figure 8–1   The mass-radius relation is shown for a white dwarf. The dashed line would apply if inverse beta decay could be neglected.

The observed properties of white dwarfs are uncertain for a number of reasons. For example, the measurements give the color temperature and not the theoretically predicted effective temperature. Line blanketing and frequency-dependent continuum absorption make the color corrections that are necessary to transform color temperature into effective temperature complicated. This is especially true for the many white dwarfs whose atmospheres are hydrogen deficient. It would be of considerable importance to have a reliable measurement of the mass-radius relation. Three white dwarfs (Sirius B, 40 Eri B, and Procyon B) are in binary systems whose orbits are known and consequently their masses can be estimated directly. The measured masses of 40 Eri B and Sirius B are 0.45 $M_\odot$ and 1.1 $M_\odot$, respectively. These results are considered reliable. The result for Procyon B ($\simeq 0.6$ $M_\odot$) is somewhat less well determined. Although the mass-radius relation agrees reasonably well with the observations, the observed radii may be somewhat larger than predicted theoretically.

The work necessary to remove one photon of frequency v_0 from the surface of a star is

$$\frac{GM}{R}\frac{hv_0}{c^2}.\tag{8–1}$$

Therefore, the energy of a photon reaching an observer at infinity is

$$hv = hv_0 - \left(\frac{GM}{R}\right)\frac{hv_0}{c^2}.\tag{8–2}$$

Equation (8–2) implies

$$\frac{v_0 - v}{v_0} = \frac{GM}{Rc^2}.\tag{8–3}$$

The gravitational fields on the surfaces of white dwarfs are sufficiently high that the gravitational redshift of spectral features is significant as compared with the usual Doppler shifts due to stellar motions. For this reason, a statistical determination of the white dwarf mass can be made. Gravitational redshift measurements yield a mean white dwarf mass that is somewhat higher than estimates based on distance determinations. However, the difference between estimates of mass based on redshift measurements and distance determinations may not be significant.

The formation of white dwarfs is easily understood on the basis of stellar evolution. After a star has exhausted its nuclear fuel, it will contract into the white dwarf state if its mass is sufficiently low. Most stars that reach the white dwarf state must have lost significant amounts of mass.

8-2 PRESSURE IONIZATION

Ions and electrons recombine in ordinary matter when it is cooled sufficiently, and consequently the equation of state is very complicated due to Coulomb interaction between atoms. However, at very high densities such as are encountered inside white dwarfs, pressure ionization prevents recombination even at zero temperature, and the pressure is primarily the result of a degenerate electron gas.

To determine an approximate condition for pressure ionization, we must first estimate the size of an atom. It follows from the uncertainty principle and the Pauli exclusion principle that the effective radius of a single electron bound to a nucleus of charge Z and radius R is approximately

$$\lambda \simeq \frac{2\pi R}{Z^{1/3}}. \tag{8-4}$$

The total energy (kinetic and potential) per electron is approximately

$$E = \frac{p^2}{2m} - \frac{Ze^2}{R} \tag{8-5}$$

where $p = h/\lambda$. Because the atom is an equilibrium state, its radius must be such that E is a minimum. The condition $\partial E/\partial R = 0$ implies

$$R \simeq \frac{Z^{-1/3}\hbar^2}{me^2} = Z^{-1/3}a_0 \tag{8-6}$$

where a_0 is the Bohr radius. If R is large as compared to the mean distance between ions, which is

$$R_i = Z^{1/3}\left(\frac{4\pi n_e}{3}\right)^{-1/3} \tag{8-7}$$

pressure ionization is expected. The condition for pressure ionization becomes

$$\frac{R_i}{R} \simeq Z^{2/3}r_e < 1 \tag{8-8}$$

where r_e is the interelectron distance in units of a_0. The above condition for pressure ionization is satisfied over most of the mass of a white dwarf, and consequently on an atomic scale the electron gas of such a configuration will be of nearly uniform density.

8-3   THE WHITE DWARF MASS LIMIT

It follows from Equation (2–31) that at zero temperature, all electron states must be occupied up to some limiting momentum p_F, and consequently the number of occupied electron states per unit volume is

$$n_e = 2 \int_0^{p_F} \frac{4\pi p^2}{h^3} \, dp = \frac{8\pi}{3} \frac{p_F{}^3}{h^3} \tag{8–9}$$

where the factor of 2 arises because the electron has two spin states. The energy density of a nonrelativistic electron gas is

$$U_e = 2 \int_0^{p_F} \frac{p^2}{2m} \frac{4\pi p^2}{h^3} \, dp = \frac{4\pi}{5} \frac{p_F{}^5}{mh^3}. \tag{8–10}$$

The corresponding electron pressure, which is the rate of transfer of momentum per unit area, is

$$P_e = 4\pi \int_0^\pi \int_0^{p_F} \frac{p_z v_z}{h^3} \sin\theta \, d\theta p^2 \, dp \tag{8–11}$$

where $v_z = p_z/m$. Equations (8–9), (8–10), and (8–11) imply

$$P_e = \frac{h^2}{5m} \left(\frac{3}{8\pi}\right)^{2/3} n_e^{5/3} \equiv K_1 \left(\frac{\rho}{\mu_e}\right)^{5/3}. \tag{8–12}$$

The electrons in a very dense ($\gtrsim 10^6$ g/cm^3) electron gas will be relativistically degenerate, and consequently $v_z = cp_z/p$. For such an electron gas, we find

$$P_e = \frac{2\pi c}{3h^3} p_F{}^4 \equiv K_2 \left(\frac{\rho}{\mu_e}\right)^{4/3} = \frac{1}{3} U_e. \tag{8–13}$$

In a white dwarf the kinetic energy of an ion ($kT \sim 10$ keV) is in general much less than the Fermi energy of an electron ($E_F \sim 1$ MeV). For this reason and also because the number density of ions is smaller by a factor of Z than that of electrons, the influence of the ions in determining the pressure is small. On the other hand, the contribution of the Coulomb interaction to the total pressure inside a white dwarf is significant. To first approximation, the Coulomb potential varies inversely as the distance between charges, and the Coulomb pressure varies as $\rho^{4/3}$. This dimensional relation shows that the pressure due to Coulomb forces will dominate at relatively low densities such as are encountered inside planets and approach an approximately constant ratio of the total pressure at very high densities ($\gtrsim 10^6$ g/cm^3) when the electron gas is relativistically degenerate. However, Coulomb effects never exceed ~ 5 percent of the total pressure inside a white dwarf.

The structure of a white dwarf of given mass and chemical composition is determined by the equation of state, $P = P(\rho)$, and the usual equations of mass conservation and hydrostatic equilibrium. These equations yield a definite relation between the mass and radius of a white dwarf. Moreover, it is found that even if the central density were infinite, the total mass of a white dwarf would be finite. This result implies that a spherically symmetric, zero temperature star more massive than a limiting value cannot exist in hydrostatic equilibrium. Numerical calculations show that this limiting mass, which is called the Chandrasekhar mass limit, is

$$M_c = \frac{1.44 \; M\odot}{(\mu_e/2)^2} \qquad (8\text{-}14)$$

for white dwarfs whose electron molecular weight μ_e is constant.

The existence of the Chandrasekhar mass limit can be understood by means of dimensional analysis. Hydrostatic equilibrium implies

$$\frac{P}{R} \sim \rho \, \frac{GM}{R^2} \sim \frac{M^2}{R^5} . \qquad (8\text{-}15)$$

If the electron gas is nonrelativistic, we have from Equation (8–12)

$$P \sim \frac{M^{5/3}}{R^5} . \qquad (8\text{-}16)$$

If this equation of state were correct at high densities, a stable white dwarf configuration of small radius could be found for any mass. However, an electron gas is relativistic at very high density, and consequently from Equation (8–13), we find

$$P \sim \frac{M^{4/3}}{R^4} . \qquad (8\text{-}17)$$

Equations (8–15) and (8–17) show that gravitational forces will dominate electron pressure for sufficiently high mass, and consequently no stable configuration can exist.

An approximate value for the mass limit can be obtained by means of physical arguments that are somewhat more illuminating than those presented above. When matter is compacted to high densities, the average Fermi energy per electron becomes higher than the electron rest energy, and consequently we can write $E_F \sim p_F c$. If N relativistically degenerate electrons are contained inside a region of radius R, the Pauli exclusion principle implies that the characteristic radius of a single electron is $\sim R/N^{1/3}$. The corresponding electron Fermi energy is

$$E_F \sim p_F c \sim \hbar \, \frac{N^{1/3} c}{R} . \qquad (8\text{-}18)$$

If $\mu_e = 2$, the gravitational energy per electron is

$$\sim - \frac{2GNm_p^{\,2}}{R}. \tag{8-19}$$

An estimate for the limiting mass is obtained by equating the above expressions for the gravitational and compressional energies. We find

$$N \sim \left(\frac{\hbar c}{G}\right)^{3/2} \frac{1}{m_p^{\,3}} \sim 10^{57} \tag{8-20}$$

from which it follows that

$$M_c \sim 2Nm_p$$
$$\sim 2 \text{ M}\odot \tag{8-21}$$

which is close to the more exact result [Equation (8–14)] obtained by numerical integration. Relativistic particles such as photons or relativistically degenerate electrons will contribute no net binding energy to a star since $\gamma = \frac{4}{3}$ [see Equation (3–10)]. For this reason, the binding energy of a white dwarf would approach a finite limit of $\simeq 7 \times 10^{50}$ ergs, which is approximately the rest mass energy of the electrons, even if the density became arbitrarily high and the electron degenerate equation of state remained physically correct.

If the Fermi energy of the degenerate electrons becomes sufficiently high, inverse beta decay,

$$e^- + (Z, A) \to (Z - 1, A) + \nu \tag{8-22}$$

will increase the mean electron molecular weight and thereby reduce the limiting mass. Most white dwarfs are likely to consist of heavy elements for which $\mu_e = 2$ such as C^{12} and O^{16}. Inverse beta decay will become significant for C^{12} if $\rho \gtrsim 2 \times 10^{10}$ g-cm^{-3}. At very high densities ($\gtrsim 4 \times 10^{12}$ g-cm^{-3}), the Fermi energy of the electrons is sufficiently high for electrons to interact directly with nucleons and thereby shift the equilibrium from heavy elements to free nucleons (mostly neutrons). Electron capture under such conditions of very high density may lead to the formation of neutron stars (see Chapter 11).

8-4 CRYSTALLIZATION AND COOLING OF WHITE DWARFS

The nuclei in ordinary matter are screened by the surrounding electron clouds, and the interaction of these electron clouds gives rise to the crystalline structure. On the other hand, electrons in very dense matter are not effective in screening nuclei, and consequently nuclei will repel each other by means of their Coulomb interactions. This circumstance means that it is energetically favorable for nuclei to arrange themselves in a regular lattice so long as the zero point energy of the ions is sufficiently small. The ions inside a cool white dwarf will form a body-centered cubic lattice. For a simple cubic lattice, the ionic distance is

$$a = \left(\frac{Z}{n_e}\right)^{1/3} \tag{8-23}$$

where n_e is the number density of electrons, and Z the charge of the ion. A reasonably good approximation to the Coulomb interaction can be obtained on the basis of the following physical model. A lattice cell is replaced by a sphere of the same volume so that

$$\frac{1}{n_i} = \frac{Z}{n_e} = \frac{4}{3}\pi r_i^3 \tag{8-24}$$

where n_i is the number of ions per cubic centimeter, and r_i is the radius of the sphere. If the Z electrons are assumed to be uniformly distributed throughout the sphere, the energy of interaction with the ion is

$$-\int_0^{r_i} \frac{Ze^2}{r} n_e 4\pi r^2 \, dr = -\frac{3}{2}\frac{(Ze)^2}{r_i} \tag{8-25}$$

where n_e is the number density of electrons. Next we must calculate the Coulomb energy due to interactions among the electrons. Since the potential inside a uniformly charged sphere of radius r_i is

$$\phi(r) = \frac{Ze}{r_i}\left(\frac{3}{2} - \frac{r^2}{r_i^2}\right) \tag{8-26}$$

the interaction energy between electrons becomes

$$\frac{1}{2}\int_0^{r_i} n_e e\phi(r)4\pi r^2 \, dr = \frac{3}{5}\frac{(Ze)^2}{r_i} \tag{8-27}$$

and consequently the total energy is

$$-\frac{9}{10}\frac{(Ze)^2}{r_i} \simeq -Z^{5/3}e^2 n_e^{1/3}. \tag{8-28}$$

The uncertainty principle implies that there is a zero point kinetic energy associated with an ion that is spatially localized in a lattice. The classical Hamiltonian of the crystal is

$$H = \sum_{i=1}^{3N} \frac{p_i^2}{2M} + V(\xi_1, \xi_2, \ldots, \xi_{3N}) \tag{8-29}$$

where the ξ_i's are the displacements of the N ions from their equilibrium positions; $p_i = M\dot{\xi}_i$ $(i = 1, 2, 3, \ldots, 3N)$ are the momenta; and V is the potential energy caused by the forces between the ions. The potential function

V is defined to be zero for the equilibrium configuration. Since the force on each ion is zero in the equilibrium state, we have

$$\left(\frac{\partial V}{\partial \xi_i}\right)_0 = 0 \qquad (i = 1, \dots, 3N) \tag{8-30}$$

where the derivatives of V are evaluated at the equilibrium states. The above arguments imply that the Hamiltonian defined in Equation (8–29) becomes

$$H = \sum_i \frac{p_i^2}{2M} + \frac{1}{2} \sum_{i,k} V_{ik}\xi_i\xi_k. \tag{8-31}$$

The classical equations of motion for the vibration of the ions are

$$M\ddot{\xi}_i = -\frac{\partial V}{\partial \xi_i} = -\sum_k V_{ik}\xi_k \tag{8-32}$$

for all i. We assume that $V_{ik} = 0$ except for $i = k$ and that V is given by Equation (8–28). Taking solutions to Equation (8–32) of the form $\xi = \xi_0 \cos \omega t$, we find

$$\omega \simeq \left(\frac{Z^2 e^2}{M r_i^3}\right)^{1/2} \tag{8-33}$$

for the frequency of vibration of the displaced ion. Equation (8–33) implies that the zero point kinetic energy per ion is

$$\frac{3}{2}\hbar\omega \simeq \frac{3}{2}\frac{\hbar Z e}{(M r_i^3)^{1/2}}. \tag{8-34}$$

The existence of a zero point kinetic energy for localized ions means that nuclear reactions can become significant even at zero temperature if the density is sufficiently high. Such nuclear reactions, which are called pycronuclear reactions, make it impossible for hydrogen to exist for an extended period of time in the interior of a white dwarf. Because of its low mass, hydrogen has a relatively high zero point kinetic energy as compared with other ions at the same density and consequently will undergo pycronuclear reactions at the relatively low density of $\simeq 10^4$ g/cm^3.

Crystalline structures will melt if their temperature is increased until the vibrations of the ions attain large amplitudes. The melting point is determined primarily by the ratio

$$\Gamma = \frac{1}{kT}\frac{(Ze)^2}{a}. \tag{8-35}$$

If the condition $\Gamma \ll 1$ holds, Coulomb interaction will be ineffective in producing a correlation between the ions. For $\Gamma \gg 1$, Coulomb forces will dominate thermal effects, and the ions will form a lattice. It has been shown that crystallization is likely to arise when $\Gamma \gtrsim 10^2$.

The heat capacity per unit volume of an ionized, nondegenerate ideal gas taken at constant volume is

$$c_v = \left(\frac{dU}{dT}\right)_v = \frac{3}{2}(n_i + n_e)k \tag{8-36}$$

where U is the internal energy, and n_i and n_e are the number densities of the ions and electrons, respectively. Equation (8–36) follows directly from the equipartition of energy which assigns $\frac{3}{2}kT$ of energy to each particle in the nondegenerate gas. As we will show, the heat capacity of a degenerate electron gas such as exists in the interior of a white dwarf is very small. The average energy of a nonrelativistic degenerate electron at zero temperature is

$$\frac{2\int_0^{p_F} E 4\pi p^2 \, dp}{(8\pi/3)\,p_F{}^3} = \frac{3}{5}E_F. \tag{8-37}$$

At some finite temperature only those electrons near the Fermi energy E_F will be excited by means of collision with the ions whose energy is $\simeq kT$. Therefore, the fraction of degenerate electrons that can be excited is $\simeq kT/E_F$. The internal energy per unit volume of the degenerate electron gas becomes

$$U = \tfrac{3}{5}n_e E_F + K n_e \left(\frac{kT}{E_F}\right) kT \tag{8-38}$$

where K is a constant which more exact calculations show to be $\pi^2/4$. From Equation (8–38) the electron heat capacity per unit volume becomes

$$c_v = \frac{\pi^2}{2} n_e k \left(\frac{kT}{E_F}\right) \tag{8-39}$$

which is small because $kT \ll E_F$.

It is well known from classical mechanics that a system of oscillators, whose physical behavior is described by a Hamiltonian such as that of Equation (8–31), possesses normal modes of oscillation. When expressed as a function of the normal coordinates Q_n and P_n, the Hamiltonian transforms into a sum of squared terms, that is,

$$H = \frac{1}{2}\sum_{n=1}^{3N}\left(\frac{P_n{}^2}{M} + M\omega_n{}^2 Q_n{}^2\right) \tag{8-40}$$

with $P_n = M\dot{Q}_n$. Since each of the $6N$ squared terms in Equation (8–40) contributes $\frac{1}{2}kT$ to the internal energy, the heat capacity per unit volume of the ions in the crystal is

$$c_v = 3n_i k. \tag{8-41}$$

Equation (8–41) remains true so long as the temperature is sufficiently high that quantum effects are unimportant. It is known from quantum theory that an oscillator of frequency v can only have integer values of $\frac{1}{2}hv$. At very low temperatures when $kT \lesssim hv$, the Boltzmann factor $\exp(-hv/kT)$ is small, and therefore an oscillator will vibrate only in the lowest energy state. It follows that since the energy of vibration does not change with temperature, the heat capacity is zero. The Debye theory of heat capacity shows that $c_v \propto T^3$ as T approaches zero. Although the classical value for the heat capacity is likely to be relevant for most observed white dwarfs, quantum effects may become important for cool, massive white dwarfs, and consequently their cooling time scales may be drastically reduced. Local order between ions will be present in white dwarf matter even before crystallization. At this point the white dwarf matter is a liquid. It can be shown that under such physical conditions, the heat capacity of the ions in a white dwarf is more nearly that of crystal than an ideal gas. For this reason, Equation (8–41) is relevant.

As a white dwarf cools toward the black dwarf state, it will contract slightly and thereby release gravitational energy. However, for small deviations from the black dwarf final state, this gravitational energy release will be absorbed as electron exclusion energy and to a lesser extent by electrostatic potential energy.

To show that this is the case, we write

$$\Delta E(\rho, T) = \Delta^T(U + V) + \Delta^\rho(U + V + \Omega) \qquad (8\text{–}42)$$

where ΔE is the energy difference (assumed small) between the white dwarf and black dwarf states. U, V, and Ω are the kinetic Coulomb and gravitational energies, respectively, of the white dwarf state. $\Delta^T U$ and $\Delta^T V$ are the changes in kinetic energy and Coulomb interaction energy, respectively, between white dwarf and black dwarf with ρ held constant. $\Delta^\rho U$, $\Delta^\rho V$, and $\Delta^\rho \Omega$ are the corresponding changes in U, V, and Ω with T held constant. We recall that $\Omega = \Omega(\rho)$ and note that vibrations make V a function of T as well as ρ.

If the radius r is changed by Δr, we have

$$\Delta^\rho \Omega = \int_0^M \frac{GM_r}{r^2} \, \Delta r \, dM_r$$

$$= -\int_0^M \frac{dP}{dM_r} (4\pi r^2 \, \Delta r) \, dM_r \qquad (8\text{–}43)$$

$$= \int_0^M P \, \Delta\left(\frac{1}{\rho}\right) dM_r$$

$$= -\Delta^\rho(U + V).$$

The last step in the above derivation follows because at zero temperature, the pressure can be written

$$P = -\frac{\partial((u + v)/\rho)}{\partial(1/\rho)} \qquad (8\text{–}44)$$

where u and v are the kinetic and Coulomb energy densities, respectively. Equation (8–43) implies that Equation (8–42) reduces to

$$\Delta E = \Delta^T(U + V). \tag{8–45}$$

It follows immediately that

$$\frac{dE}{dT} = \int dV c_v(T) \tag{8–46}$$

where $c_v(T)$ is the heat capacity per unit volume taken at constant volume. Equation (8–46) shows that a white dwarf that is sufficiently close to the black dwarf state cools at the expense of thermal energy (see Figure 8–2). To first order the decrease in gravitational energy is equal to the increase in electron exclusion energy and Coulomb interaction energy.

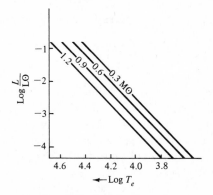

Figure 8–2   Cooling curves for white dwarfs of varying mass.

The thermal conductivity due to electron conduction in the core of a white dwarf is very high, and consequently the core is nearly isothermal. This means that the temperature gradient occurs primarily in the thin, mostly semidegenerate surface layers. We assume that the outer layers are in radiative equilibrium and that Kramers' opacity law ($\kappa = \kappa_0 \rho T^{-3.5}$) is approximately correct near the surface. Integration of the equations of hydrostatic and radiative equilibrium through the envelope of the white dwarf gives

$$\rho = \left(\frac{32\pi\mu acm_pGM}{25.5k\kappa_0 L}\right)^{1/2} T^{3.25} \tag{8–47}$$

where it has been assumed that the mass of the envelope is small. An approximate relation between the density and temperature in the transition region between the degenerate core and the nondegenerate surface layers can be found if we set the electron pressure that would result if the electrons were nondegenerate equal to that which would result if they were degenerate. We find

$$\frac{\rho_s k T_{core}}{\mu_e m_p} = K_1 \left(\frac{\rho_s}{\mu_e}\right)^{5/3}. \tag{8-48}$$

Equation (8-48) implies

$$\rho_s = \mu_e \left(\frac{k T_{core}}{m_p K_1}\right)^{3/2}. \tag{8-49}$$

If we substitute the expression for the density in the transition region equal to ρ_s in Equation (8-47), we find

$$L = KM T_{core}^{3.5} \tag{8-50}$$

where

$$K = \frac{32\pi\mu a c m_p^4 G K_1^3}{25.5 k^4 \kappa_0 \mu_e^2}.$$

If thermal energy is the only energy source in the interior of a white dwarf, L can be written

$$L = -\frac{d}{dt} \int \int c_v \, dT \, dV. \tag{8-51}$$

For most white dwarfs, $c_v = 3n_i k$ throughout the core, which contains nearly all the thermal energy. This circumstance allows us to use Equations (8-50) and (8-51) to find both L and T_{core} as a function of time. Recent calculations predict the presence of convective regions caused by partial ionization in the outer layers of relatively cool white dwarfs. Such convective zones may drastically reduce their lifetimes.

In Chapter 2 we showed that a main-sequence star is secularly stable because its effective heat capacity is negative (that is, its interior temperature increases when thermal energy is lost). On the other hand, the effective heat capacity of a white dwarf is positive since it cools as it contracts. For this reason, a white dwarf would be secularly unstable if the luminosity were due to a nuclear fuel such that $\varepsilon \propto T^\nu$ with $\nu > 3.5$ since the temperature sensitivity of the energy gains would be greater than the temperature sensitivity of the energy losses. It can also be shown that small quantities of nuclear burning fuel in the interior of a white dwarf would lead to pulsational instability. Evidence for instability is not observed in normal white dwarfs. This result provides further evidence that nuclear energy generation is not responsible for their observed luminosity.

8-5 RAPIDLY ROTATING WHITE DWARFS

In deriving the Chandrasekhar mass limit, we neglected rotation and assumed that the white dwarf was spherically symmetric. If a star conserved much of its main-sequence angular momentum, its rotational energy in the white dwarf state would become very large. Although uniform rotation cannot appreciably increase the mass limit, it has been shown that one can construct differentially rotating white dwarf models with masses much larger ($\gtrsim 4\,M_\odot$) than the Chandrasekhar mass limit. The existence of such rapidly rotating white dwarfs has not been demonstrated observationally. However, such objects might be difficult to distinguish observationally from white dwarfs with lower mass and very broadened spectral features.

The lifetime of a rapidly rotating white dwarf is determined by the circumstance that the ionic component would crystallize in the absence of differential rotation. Since a differentially rotating white dwarf cannot cool below the temperature of crystallization, the luminosity will be relatively high and consequently the lifetime much shorter than that of the galaxy.

Magnetic Fields

The recent discovery of circularly polarized light from a white dwarf provides indirect evidence for a surface magnetic field of about 10^7 G. The presence of such intense surface magnetic fields makes it likely that white dwarfs emit nonthermal radiation. It has been suggested that rapidly rotating white dwarfs with strong magnetic fields may be sources of cosmic ray particles.

Many white dwarfs are observed to have narrow spectral features such as the Balmer series of hydrogen. The observed widths of these spectral features show that the surface magnetic fields of these objects are $\lesssim 10^5$ G. On the other hand, there exists a class of white dwarfs that show no evidence of narrow spectral features. Such objects are possible candidates for high surface fields and rapid rotation.

CHAPTER 9

Supernovae and the Formation of Heavy Elements

9-1 OBSERVED PROPERTIES OF SUPERNOVAE

A supernova is an exploding star that ejects a large fraction of its mass at very high velocities. At maximum, its observed luminosity approaches that of an entire galaxy. Two major classes of supernovae, called type I and type II, are distinguished. Typical supernova light curves are shown schematically in Figure 9–1. Type II supernovae appear only in spiral galaxies, and consequently it is likely that they are associated with massive stars. On the other hand, type I supernovae occur with approximately equal rates in both spiral and elliptical galaxies. Since the largest elliptical galaxies are believed to contain approximately ten times as many stars as the largest spiral galaxies, the rate of type I supernova production *per star* is substantially greater in spiral galaxies than elliptical galaxies.

Supernovae are of great interest to astronomers and astrophysicists because they represent the most spectacular stellar events. The supernova outburst involves an almost instantaneous release of energy that is in excess of the binding energy of a star (that is, $> 10^{50}$ ergs), and consequently it is widely believed that matter attains unusual physical conditions during and immediately after a supernova outburst. The production of high-velocity stars, nonthermal radiation (mostly synchrotron radiation), cosmic radiation, heavy elements, and probably gravitational radiation are associated with supernova outbursts. Moreover, the discovery of pulsars (see Chapter 10) has provided convincing evidence that neutron stars (see Chapter 11) are the remnants of many supernovae.

At maximum light, the absolute visual magnitude of a type I supernova is $M_v \sim -19$, (10^9-10^{10} L$\odot$). The corresponding optical luminosity of a type II supernova is about a magnitude less. These very high luminosities are much greater than those of novae, which range from about $M_v = -6$ to $M_v = -9$, and consequently it is evident that novae and supernovae are unrelated events. The integrated visual output is about 10^{49} ergs for a

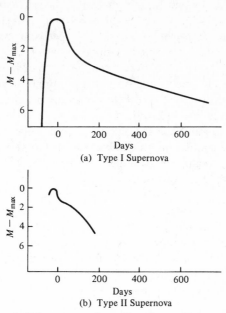

(a) Type I Supernova

(b) Type II Supernova

Figure 9–1   Schematic light curves of type I and type II supernovae.

type I supernova and approximately ten times less for a supernova of type II. Although the total energy released in a supernova is unknown, since not all forms of energy are observable, the observations indicate that it is $\gtrsim 10^{50}$–10^{51} ergs. Such a large explosive release of energy is sufficient to disrupt the entire star. For this reason, the likely end products of a supernova are neutron stars, collapsed objects, very hot white dwarfs, or no remnant.

Unfortunately, the data concerning the occurrences of supernovae are inadequate to obtain an accurate estimate of the production rate per galaxy. However, it is clear that it is significantly less than the predicted formation rate of white dwarfs (~ 3 per year). The best estimate for the rate of supernovae is ~ 1 every 100 yr per spiral galaxy for type II supernovae, and ~ 1 every 300 yr per galaxy (spiral or elliptical) for type I supernovae.

The light curves of type I supernovae show a characteristic linear decline (one magnitude per 50–100 days) with no secondary maxima such as are observed for novae. The observed linear decline of the type I supernova is the key observation that theories of the light curve must explain. The spectra of different type I supernovae are very similar in appearance. They all show broad features that are observed before maximum light. The nature of the spectral features is not fully understood. There is some debate as to whether they are broadened emission lines or a combination of absorption and emission lines. The short wavelength spectra (<5000 Å) maintain their general structure long after maximum light. A particularly prominent feature, which may be caused by emission from singly ionized helium, is observed at 4600 Å. Although the short wavelength spectrum is largely stable, certain features associated with the longer wavelength spectrum vanish and appear in an irregular manner reminiscent of ordinary novae.

The spectra of type II supernovae resemble those of novae and are very different from those of type I supernovae. A continuous spectrum with an intense UV component is observed before maximum light and for a short time thereafter. The color temperature is that of an O star at maximum light at which time emission and absorption bands are not observed. After maximum light, the blue continuum gradually declines and very broad emission and absorption features appear. Although the spectrum resembles that of an ordinary nova, forbidden transitions appear much later, and the observed linewidths are much greater (5–10×10^3 km/sec). Since the appearance of forbidden lines depends primarily on the density in the expanding supernova shell (it must be sufficiently low), their relatively late appearance indicates that the ejected mass in a type II supernova is at least several orders of magnitude greater than in an ordinary nova. A small fraction of type II supernovae, which are sometimes called type III supernovae, appear to have ejected envelopes of very large mass.

9–2   RUNAWAY STARS

A significant number of massive (that is, OB type) stars are known to have very high velocities (50–100 km/sec). Some of these high-velocity stars appear to be moving away from stellar associations that contain other young massive stars. Observational evidence indicates that these stars do not belong to the smooth tail of some velocity distribution. Moreover, they occur at least five times less frequently in binary systems than ordinary low-velocity OB stars. Approximately 75 percent of ordinary OB stars are members of binary systems.

It has been suggested that many high-velocity stars were previously members of binary systems in which the other member exploded as a supernova. If the mass lost by the exploding star is sufficiently great, the explosion will disrupt the binary system.

Consider a binary system such that the components move in circular orbits about their center of mass and are separated by a distance a. Their masses are M_A and M_B. In the center of mass (cm) frame, the velocities of stars A and B can readily be shown to be

$$v_A = M_B \left[\frac{G}{a(M_A + M_B)} \right]^{1/2}$$

$$v_B = M_A \left[\frac{G}{a(M_A + M_B)} \right]^{1/2},$$

(9–1)

respectively.

The mass ejected during a supernova outburst is ejected at very high velocity (~ 7000 km/sec), which is much greater than the orbital velocities of the binary components. For this reason, it is reasonable to assume that the mass loss from the exploding star M_A takes place instantaneously. We let qM_A be the mass of the remnant and $(1 - q)M_A$ the ejected mass. We assume that the mass is ejected with spherical symmetry about M_A.

The velocities of stars A and B must be continuous at the instant of the supernova outburst since the acceleration must remain finite. On the other hand, the potential and kinetic energy of the binary system is discontinuous if the mass ejection is sudden. Immediately after the explosion, the kinetic energy in the original c-m frame is

$$T = \tfrac{1}{2}qM_A v_A{}^2 + \tfrac{1}{2}M_B v_B{}^2$$

$$= \frac{1}{2}\frac{GM_A M_B}{a(M_A + M_B)}[qM_B + M_A].\tag{9-2}$$

The corresponding potential energy is

$$U = -q\,\frac{GM_A M_B}{a}.\tag{9-3}$$

Equations (9–2) and (9–3) imply that the total energy in the original c-m system is

$$E' = \frac{GM_A M_B}{2a(M_A + M_B)}[M_A(1 - 2q) - M_B q].\tag{9-4}$$

After the explosion, the total momentum in the original c-m frame is

$$p = M_B v_B - qM_A v_A.\tag{9-5}$$

Therefore, the energy associated with the motion of the c-m is

$$\frac{p^2}{2(qM_A + M_B)}.$$

Equations (9–4) and (9–5) imply that the total energy in the new c-m frame is

$$E'' = E' - \frac{p^2}{2(qM_A + M_B)} = \frac{qGM_A M_B}{2a(qM_A + M_B)}[M_A - M_B - 2qM_A].\tag{9-6}$$

Equation (9–6) shows that the binary will be disrupted by a supernova explosion if

$$M_A - M_B > 2M_n\tag{9-7}$$

where M_A is assumed greater than M_B, and M_n is the mass of the remnant.

9-3   MAGNETOBREMSSTRAHLUNG (SYNCHROTRON RADIATION)

The remnants of supernova explosions are observed to be nonthermal radio sources. Since relativistic particles including electrons are likely to be produced as a result of explosions, it is plausible that nonthermal radiation from supernova remnants is caused by the magnetobremsstrahlung of relativistic electrons.

As shown in Section 1–13, the path of an electron moving in a uniform magnetic field is that of helical motion with the gyrofrequency

$$\omega_B = \frac{eB}{mc\gamma}$$

$$\gamma = \frac{1}{\sqrt{1 - (v^2/c^2)}}.$$

(9–8)

The quantity $\omega_c = eB/mc$ is called the cyclotron frequency. Since the electron is accelerated by the magnetic field, it will radiate. In Equation (2–53) we gave an expression for the power radiated by a charge undergoing nonrelativistic motion. This expression is valid in an inertia coordinate system that has the instantaneous velocity of the electron. The energy E transforms under a Lorentz transformation like the fourth component of the four-vector,

$$P_\mu = \left(\mathbf{p}, \frac{iE}{c} \right) \qquad \mu = 1, 2, 3, 4$$

(9–9)

where $\mathbf{p}$ is the momentum of the particle. Moreover, the time t also transforms as the fourth component of a four-vector, namely,

$$(\mathbf{r}, ict)$$

(9–10)

where $\mathbf{r} = (x, y, z)$ is the position vector. It follows that since dE and dt transform in the same way, the total power radiated $-dE/dt$ must be a scalar quantity, which implies that it is invariant under Lorentz transformations. It follows that since the power radiated by an accelerated charge is the same in all inertial coordinate systems, the general formula for the total power radiated can be obtained by writing down a Lorentz invariant generalization of Equation (2–53) that reduces to that equation when the motions of the particle are nonrelativistic. Since the scalar product of two four-vectors is a scalar quantity, the general expression for the total power radiated becomes

$$-\frac{dE}{dt} = \frac{2}{3} \frac{e^2}{m^2 c^3} \left(\frac{dP_\mu}{d\tau} \right)^2 = \frac{2}{3} \frac{e^2}{m^2 c^3} \left[\left(\frac{d\mathbf{p}}{d\tau} \right)^2 - \frac{1}{c^2} \left(\frac{dE}{d\tau} \right)^2 \right]$$

(9–11)

where $d\tau = dt/\gamma$ is the proper time element, and P_μ is the energy-momentum four-vector. The expression for $-dE/dt$ given in Equation (9–11) reduces to the equation

$$- \frac{dE}{dt} = \frac{2}{3} \frac{e^2}{c^3} \gamma^4 \left[\left(\frac{d\mathbf{v}}{dt} \right)^2 - \left(\mathbf{v} \times \frac{d\mathbf{v}}{dt} \right)^2 \right] \qquad (9\text{–}12)$$

where use has been made of the relations

$$\frac{d\mathbf{p}}{dt} = \frac{d}{dt}(\gamma m \mathbf{v}) = m\gamma \frac{d\mathbf{v}}{dt} + \frac{m\mathbf{v}}{2c^2} \gamma^{3/2} \frac{dv^2}{dt} \qquad (9\text{–}13)$$

and

$$\mathbf{p} = \frac{\mathbf{v}}{c^2} E.$$

It follows from Equation (1–137) that for motion in a magnetic field, we have

$$\frac{d\mathbf{v}}{dt} = \frac{e}{\gamma mc} (\mathbf{v} \times \mathbf{B}). \qquad (9\text{–}14)$$

Substituting Equation (9–14) into Equation (9–12), we find

$$- \frac{dE}{dt} = \frac{2}{3} \frac{e^4 B_\perp^2}{m^2 c^3} \left[\left(\frac{E}{mc^2} \right)^2 - 1 \right]. \qquad (9\text{–}15)$$

For an ultrarelativistic electron, Equation (9–15) reduces to

$$- \frac{dE}{dt} = \sigma_T c \frac{B_\perp^2}{4\pi} \gamma^2 = 1.6 \times 10^{-15} B_\perp^2 \gamma^2 \qquad (9\text{–}16)$$

where the Thomson cross section σ_T is defined in Equation (2–58). Equation (9–15) shows that protons of energy E emit synchrotron radiation at a rate $(m/m_p)^4 \sim 10^{13}$ times less than electrons with the same energy. For this reason, only relativistic electrons are likely to emit significant synchrotron radiation unless the magnetic fields are very large. The above description of synchrotron radiation is correct as long as

$$h\nu \ll E \qquad (9\text{–}17)$$

where ν is the frequency of the emitted radiation.

The spectrum of synchrotron radiation can be qualitatively understood by means of the following physical arguments. Imagine a relativistic electron that is in circular motion in a magnetic field. An observer at large distances from the electron will measure pulses of radiation at time intervals given by

$$\tau = \frac{2\pi}{\omega_B}. \qquad (9\text{–}18)$$

The spectrum of the pulses will consist of harmonics of ω_B. Since the radiation is emitted within a cone of angular width, $\theta \sim mc^2/E = 1/\gamma$, along the direction of the particle's motion, the power spectrum of the emitted radiation will extend to angular frequencies

$$\omega_{max} = \omega_B \left(\frac{E}{mc^2}\right)^3 = \frac{eB}{mc}\left(\frac{E}{mc^2}\right)^2 \qquad (9\text{-}19)$$

where one factor of γ is due to the beamwidth, and the remaining two are a consequence of the Doppler effect. The calculated spectral distribution of a single electron in circular motion is given in Figure 9-2.

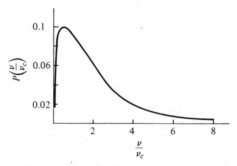

Figure 9-2   Power radiated per unit frequency interval of relativistic electron orbiting in magnetic field.

The radio emission from a nonthermal radio source such as a supernova remnant or a radio galaxy is the result of emission from a distribution of relativistic electrons moving in magnetic fields. It is plausible to assume a power law energy spectrum for these particles. If the relativistic electrons are assumed to be distributed uniformly throughout the source, we have

$$N(E)\, dE = KE^{-\gamma}\, dE \qquad (9\text{-}20)$$

for the number of relativistic electrons per unit volume in the energy range E, $E + dE$. We assume that the magnetic field is uniform throughout the source and that $B_\perp{}^2 = \frac{2}{3}B^2$.

An approximate expression for the intensity of radiation I_ν can be found by assuming that all the radiation from electrons of energy E is radiated at a frequency $\nu = 0.29\nu_c$ (see Figure 9-2). We have

$$I_\nu = \frac{L}{4\pi}\int\left(-\frac{dE}{dt}\right)\delta(\nu - 0.29\,\nu_c)KE^{-\gamma}\, dE$$

$$= A(\gamma)\frac{e^3}{mc^2}\left(\frac{3e}{4\pi m^3 c^5}\right)^{(\gamma-1)/2} B^{(\gamma+1)/2}KL\nu^{-(\gamma-1)/2} \qquad (9\text{-}21)$$

where $A(\gamma) = 0.31 \times (0.24)^{(\gamma-1)/2}$; L is the dimension of the source; and $-dE/dt$ is given in Equation (9–16). The quantity $\alpha = (\gamma - 1)/2$, which is called the spectral index, is approximately 0.8 for most supernova remnants.

9–4   REMNANTS OF TYPE II SUPERNOVAE

The remnants of type II supernovae are filamentary nebulae that are concentrated toward the galactic plane. These nebulae are identified with nonthermal radio sources and are thereby distinguished from other nebulae with filaments. The spectral indices of these remnants are typically $\alpha = 0.8$. The Cygnus loop is an example of a relatively old ($\sim 30,000$ yr) type II supernova that exploded close to the solar system. Its relatively slow expansion velocity indicates that it has been slowed down by the interstellar medium. Therefore, its filamentary structure was likely produced by a high-velocity expanding supernova shell that compressed the surrounding interstellar medium.

Cassiopeia A, which is the strongest celestial radio source, is a type II supernova. Its young age (~ 300 yr old) indicates that its observed expansion velocity (~ 7000 km/sec) is nearly equal to its initial expansion velocity. The radio emission from Cassiopeia A is decreasing by approximately 1 percent per year. If the magnetic flux is conserved during adiabatic expansion, the magnetic field strength and particle energies will decrease as the shell expands, and consequently the intensity of the synchrotron radiation will decrease. It should be emphasized that the discovery of pulsars shows that active energy sources exist at the sites of supernova outbursts, and therefore the continued injection of relativistic particles and magnetic flux is anticipated for many, perhaps most, supernovae.

The observed optical spectra of Cassiopeia A indicates that the remnant consists of a system of approximately 100 rapidly moving "knots" of gas and a system of nearly stationary bright regions observed in the emission of individual spectral lines. The differences in velocity between different high-velocity clouds suggests that some of these massive clouds are still being accelerated. The system of bright regions is likely to be the result of the interaction of the supernova shell with interstellar gas clouds or previously ejected mass from the presupernova star itself.

9–5   CRAB NEBULA

The Crab nebula, which is a remnant of a supernova explosion observed by Chinese astronomers in the year 1054 A.D., has played a unique role in the development of modern astrophysics. Before the discovery that its radio emission and much of its optical emission are the result of synchrotron radiation, thermal radiation was the only type of radiation generally discussed by astronomers. The Crab nebula is known to be a strong x-ray source. The discovery of hard x rays from the nebula indicates the presence of very-high-energy electrons ($>10^{13}$ eV) and therefore suggests that the Crab is likely to be an important source of cosmic ray particles. More recently, a pulsar has been identified with a star near the center of the nebula. The presence of this pulsar, which is likely to be a rotating neutron star, shows that the present Crab nebula contains an active energy source and therefore is not merely the relic of a past explosion.

The spectrum of the Crab nebula resembles that of planetary nebulae. The strongest emission lines are the usual nebula emission lines of [OII], [OIII], [NII], [SII], and the Balmer series of hydrogen. However, although ~ 90 percent of the visual radiation from planetary nebulae is in the form of emission lines and only ~ 10 percent is in the form of continuous radiation, the continuous component from the Crab nebula constitutes ~ 80 percent of its visual radiation. Moreover, the simultaneous presence of ions with very different ionization potentials such as [OI], [OII], and [OIII] shows that the excitation of these transitions is not primarily by means of blackbody radiation. With the exception of the He/H ratio, which appears to be higher than normal by a factor of ~ 3, the estimated abundances in the Crab nebula are approximately normal. This is surprising in view of the very unusual nature of this object. The absence of [Fe] transitions is particularly surprising since Fe is expected to be produced in supernova explosions and also be present in the surface layers of neutron stars.

Photographs of the Crab nebula that are taken with narrow-band filters centered on various emission lines show that the emission lines arise from a system of filaments that surround the rest of the nebula. On the other hand, the continuous radiation comes from the entire volume, which is called the "amorphous mass," although it is itself highly structured.

The filamentary system has a nearly regular elliptic form as seen projected along the line of sight. Individual filaments have been observed to undergo noticeable changes, and it is well known that the nebula is expanding. The outward motions of filaments and other features can be measured by means of a series of photographs taken over a number of years. The observed motions are primarily radially outward, and filaments close to the edge of the nebula have larger motions than those inside. The apparent centroid of the nebula, as well as the characteristic time scale for nebula expansion, can be obtained by extrapolating backward in time. The Crab pulsar has been identified as one of two stars that are near the center of convergence of the nebula.

If the observed proper motions of the nebula are assumed to be independent of time, the predicted convergence of the various elements arises at a time $\sim 10^2$ yr after the known time of the supernova outburst. This result indicates that the present observed proper motions are larger than the mean motions. If it is assumed that the inferred increase in the proper motions is the result of a constant acceleration a, we find

$$d = \tfrac{1}{2}at^2 + v_0 t$$
$$v = at + v_0 \tag{9-22}$$

where d is the semimajor axis of the projected nebula ellipse; v_0 is the initial expansion velocity; $v = 0.22$ sec/yr, and $a \sim 2 \times 10^{-5}$ sec/yr^2. If the distance to the Crab nebula is taken to be 2 kpc, a is equal to 0.0007 cm/sec^2 at the ends of the semimajor axis.

It is of considerable interest to estimate the input of kinetic energy required to produce the observed acceleration of the nebula. The change in kinetic energy is

$$\frac{d(Mv^2/2)}{dt} = Mva + \frac{v^2}{2}\frac{dM}{dt} \tag{9-23}$$

where M, the mass of the nebula, increases as the nebula expands into interstellar space. If we take $M = 1$ M$\odot$, $v = 10^3$ km/sec, and $a = 0.0007$ cm/sec^2, Equation (9–23) becomes

$$\frac{d(Mv^2/2)}{dt} = 1.4 \times 10^{38} + 1 \times 10^{38}n \text{ ergs/sec} \tag{9-24}$$

where n is the number density of atoms in interstellar space. Relativistic particles and/or the magnetic fields of the nebula could produce the observed acceleration of the Crab nebula. A magnetic field of $\sim 10^{-3}$ G or a total energy content of $\sim 10^{49}$ ergs in relativistic particles that are distributed over the volume ($\sim 10^{56}$ cm^3) of the nebula would be sufficient to cause its observed acceleration.

If the linearly polarized continuum radiation of the Crab nebula is observed with high resolution, most of the radiation is found to come from thin "fibers" that appear to be aligned with the magnetic field of the nebula, but not connected with the filaments. Therefore, it appears that the continuous optical and x-ray radiation is emitted from relatively compressed regions whose total volume is probably small as compared with that of the entire nebula.

Several oscillating wisps have been observed in the Crab nebula. The production of these wisps appears to be correlated with sudden changes in the observed period of the Crab pulsar. It is plausible to assume that these wisps are plasma waves.

If the nebula is studied with high angular resolution, good correlation between optical and radio features is observed. This important observational result indicates that the radio and optical radiation are of similar origin. Radio studies of the nebula indicate that the magnetic field strength is $\sim 10^{-4}$ G, and the density of nonrelativistic electrons ~0.05 cm^{-3}. The spectra of the Crab nebula and pulsar are shown schematically in Figure 9–3. The total radiated energy is $\sim2 \times 10^{38}$ ergs/sec. As will be argued in Chapter 10, it is plausible to assume that the rotational energy of the pulsar is the energy source for this radiation.

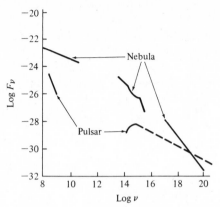

Figure 9–3   The measured flux density from crab nebula and pulsar.

9–6   MECHANISMS FOR SUPERNOVA OUTBURST

Energy Sources

Gravitational and nuclear energy release are the two available energy sources for a supernova outburst. In Chapter 8 we pointed out that the maximum binding energy of a white dwarf, which is supported by the pressure of degenerate electrons, is approximately the rest mass energy of its electrons. Since the energy density at each point in space must be positive, the total energy of a star must remain positive, and consequently the maximum binding energy of a collapsed object is approximately its total rest mass energy. This circumstance implies that the energy release by means of gravitational collapse must be less than the rest mass energy of the star. For most stars, only the inner 1.4 M$\odot$ will collapse, and so the total energy release must be $\lesssim 10^{54}$ ergs. The observed energy from supernovae is, of course, much less than this upper limit.

The fusion of hydrogen is the principal nuclear energy source in a star. Approximately 10^{52} ergs/M$\odot$ is available from hydrogen burning. However, in order that a nuclear fuel be able to produce a supernova outburst, the required energy release ($\gtrsim 10^{50}$–10^{51} ergs) must take place in less than a degenerate core dynamical time scale (that is, seconds). Such very high rates of hydrogen burning are unlikely to arise in the interiors of stars. The two modes of hydrogen burning are the proton chain and the CNO cycle.

The proton chain is not a suitable set of nuclear reactions because the first reaction

$$p + p \rightarrow D^2 + e^+ + \nu \qquad (9\text{-}25)$$

is not very temperature sensitive and consequently always very slow. The CNO cycle requires a minimum of about 10^3 sec for completion (see Section 2–7) if the temperature is $\lesssim 10^{8}°K$. Since hydrogen burning is unlikely to take place at temperatures in excess of $10^{8}°K$ inside stars, the above requirement limits the attainable rates of energy generation to about 10^{14} ergs/g-sec. These rates are too low to produce a supernova explosion. If supernova outbursts are the result of nuclear explosions, it is likely that explosive carbon and/or oxygen burning are responsible since very high burning rates are attainable by means of these reactions.

Neutrino Emission and Explosive Carbon and Oxygen Burning

The particular physical mechanism responsible for a supernova outburst is a consequence of the density, temperature, and chemical composition of the central regions of a star during its final evolutionary stages. One of the primary goals of stellar evolutionary studies is to determine the range of physical conditions that is likely to be relevant and thereby hopefully isolate the physical mechanisms responsible for supernovae. The presence of neutrino emission processes, notably

$$\gamma_{plasma} \rightarrow \nu + \bar{\nu} \quad \text{(plasma neutrino)}$$

and $\qquad\qquad\qquad\qquad\qquad\qquad\qquad\qquad\qquad\qquad\qquad\qquad (9\text{-}26)$

$$\gamma + e^- \rightarrow e^- + \nu + \bar{\nu} \quad \text{(photoneutrino)}$$

have a decisive influence on the final evolution of a star. Figure 9–4 shows evolutionary paths in the $\rho\text{-}T$ plane of selected stellar evolutionary sequences. The presence of neutrino emission processes is seen to delay the ignition of carbon and oxygen burning in stars with sufficiently low mass. Carbon and oxygen burning will not take place at all for stars less massive than ~ 1.4 M$\odot$ that initially contain significant amounts of hydrogen or helium. Carbon burning is expected to occur under conditions of very high density ($2\text{–}5 \times 10^9$ g/cm^3) in stars whose mass is <9 M$\odot$, and similarly oxygen burning is likely to arise under very degenerate conditions for stars <15 M$\odot$. Nuclear energy generation that arises under conditions of very high degeneracy is a likely mechanism for a supernova since the potentially explosive release of energy (2×10^{51} ergs) is greater than the binding energy of the star. It has been shown that for the relevant high central densities, carbon burning can produce a sufficient overpressure to eject the envelope of the star at very high velocities ($\gtrsim 10^4$ km/sec). Calculations predict that explosive carbon burning will disperse the entire star into space if the initial central density is $\lesssim 1 \times 10^{10}$ g/cm^3 but greater than about 2×10^7 g/cm^3. On the other hand, rapid cooling by means of inverse beta decay followed by beta decay (called

Urca process) is sufficient to reverse the expansion of the core and thereby lead to the formation of a remnant neutron star or black hole if the initial density is $\gtrsim 1 \times 10^{10}$ g/cm^3 or convection can increase the efficiency of the Urca reactions, which are

$$e^- + (Z, A) \rightarrow (Z - 1, A) + v$$
$$(Z - 1, A) \rightarrow (Z, A) + e^- + \bar{v}.$$

$e^- - e^+$ Pair Production Supernova

Explosive oxygen burning that is induced by $e^- - e^+$ pair formation is a possible physical mechanism for the production of supernovae. If the interior of a star evolves into a region of the ρ-T plane with very high temperature and relatively low density (see Figure 9–4), the production of $e^- - e^+$ pairs under conditions of thermal equilibrium with photons

$$e^- + e^+ \rightleftarrows \gamma + \gamma$$

will reduce

$$\Gamma \equiv \frac{\rho}{P}\left(\frac{dP}{d\rho}\right)_{s=\text{entropy}} \tag{9–27}$$

below $\frac{4}{3}$. Dynamical instability will result because the stability integral

$$\int (\Gamma - \tfrac{4}{3})P \, dV \tag{9–28}$$

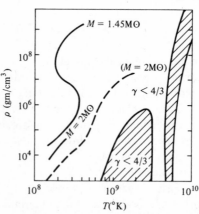

Figure 9–4   The curves represent central density as function of temperature. The solid curve labeled $M = 1.45M\odot$ shows the effect of neutrino emission on the path in the ρ-T plane of star with helium burning shell. The dashed curve labeled $M = 2M\odot$ is the corresponding path of $2M\odot$ pure oxygen star for which neutrino emission is unimportant. The early evolution of models for a $2M\odot$ star with a helium burning shell is shown for comparison. The He–Fe and pair-formation instability strips are labeled $\gamma < \frac{4}{3}$.

becomes negative. Detailed calculations show that the onset of dynamical instability causes a very rapid contraction ($\sim 0.1 \times$ gravity) that leads to a sudden increase in the interior temperature. Because of the very high temperature sensitivity of the oxygen-burning reactions ($\varepsilon_{O^{16}+O^{16}} \propto T^n$, $n \sim 30$), the induced increase in central temperature is sufficient to lead to explosive oxygen burning and thereby disperse the entire star into interstellar space.

The above implosive process that precedes the nuclear explosion arises because the pressure per unit mass energy is greater for relativistic matter such as photons than for $e^- - e^+$ pairs under the relevant physical conditions. This circumstance means that the rapid transformation of photons into $e^- - e^+$ pairs can lead to a sudden reduction in pressure and thus an implosive process. It should be emphasized that this type of supernova can only arise in very massive stars and therefore must be very rare at the present epoch.

Rotational Energy Release

The above models for supernova outbursts have assumed spherical symmetry. However, the importance of rotational energy E_{rot} as compared to gravitational energy Ω increases during collapse because

$$\frac{E_{rot}}{\Omega} \propto \frac{1}{R} \tag{9-29}$$

for constant angular momentum. The observed rotational velocities of stars suggests that unless a mechanism for removing much of the angular momentum from stars is operative, collapsing stars should deviate very markedly from spherical symmetry at high (that is, $\sim$ nuclear) densities. It is likely that significant departures from spherical symmetry play a fundamental role in causing many supernova outbursts, and consequently rotational energy is a likely energy source for many supernovae.

9-7 FORMATION OF HEAVY ELEMENTS

Solar system abundances are estimated by studying the compositions of certain meteorites, notably chondritic meteorites, that are believed to be representative of nonvolatile elements and also by measuring the relative strengths of solar spectral lines. With the aid of model solar atmospheres, the abundances of various isotopes can be determined relative to hydrogen. These two methods of determining solar system abundances enable a nearly complete set of element abundances to be obtained. Nucleosynthesis is concerned with explaining the relative abundance of isotopes on the basis of known nuclear properties and plausible physical conditions that are likely to exist at the time of their formation. It is widely believed that supernovae play a dominant role in the formation of heavy elements such as iron peak elements.

The very great differences that are observed between the spectra of various stars is primarily a consequence of variations in surface gravity and effective temperature rather than intrinsic chemical composition. However, there exist a number of stars whose chemical compositions appear to be very unlike that of the sun. Physical processes on the surfaces of stars such as gravitational separation, radiation pressure, and strong magnetic fields are undoubtedly responsible for many apparent chemical peculiarities. However, the very great deficiencies in heavy elements (by a factor of 10^2–10^3) that are observed for extreme population II stars is believed to support the point of view that heavy elements have been synthesized in the interiors of stars sometime after the formation of the galaxy. The considerable spread in observed chemical compositions among old star clusters provides evidence that a significant fraction of the synthesis of heavy elements took place very soon ($\lesssim 10^9$ yr) after the formation of the galaxy.

Nuclear forces are responsible for the large nuclear binding energies, which are typically 8 MeV per nucleon. The binding energy of a nuclide of atomic weight A and charge Z can be written

$$BE = Zm_p c^2 + (A - Z)m_n c^2 - M(A, Z)c^2 \qquad (9\text{--}30)$$

where $M(A, Z)$ is the mass of the nuclide, and $N = A - Z$ is the number of neutrons.

The gross features of the binding energy of nuclei can be predicted by the semiempirical mass formula

$$M(A, Z) - Zm_p - (A - Z)m_n$$
$$= -a_1 A + a_2 A^{2/3} + a_3 Z^2 A^{-1/3} + a_4 (N - Z)^2 A^{-1} \pm a_5 A^{-1}$$
$$(9\text{--}31)$$

where the a's are positive constants. The first term on the right-hand side of Equation (9–31) is due to the nuclear interaction. It is negative because the nuclear force is an attractive force and is proportional to A because, unlike the Coulomb force, the nuclear force can have only a fixed number of bonds per nucleon. The saturation of nuclear forces implies that nuclides have a constant density like a liquid drop. The second term in Equation (9–31) is a correction to the first term. It arises because the nucleons on the surface of a nucleus do not have as many bonds as those in the interior. This term, which is similar to the surface tension in a liquid drop, is proportional to the surface area of the nucleus and consequently proportional to $A^{2/3}$.

The third term in Equation (9–31) is due to the positive electrostatic energy, which is

$$\frac{3}{5}\frac{Z^2 e^2}{R} \qquad (9\text{--}32)$$

for a uniformly charged sphere of radius R. Since the radius R of a nucleus is

$$R = 1.4 \times 10^{-13} A^{1/3} \text{ cm} \tag{9-33}$$

the positive energy of repulsion is proportional to $Z^2 A^{-1/3}$. The fourth term is the result of the exclusion principle and therefore cannot be understood on the basis of the classical liquid drop model of the nucleus. The last term is due to the tendency of nucleons to be bound in pairs. The plus sign arises if Z and N are both odd, and the minus sign arises if Z and N are both even. For other nuclei the term is zero.

Although the semiempirical mass formula is useful in predicting certain general properties such as the stability of nuclei, actual nuclei show marked variations from nucleus to nucleus, that cannot be fully understood on the basis of the semiempirical mass formula. If the binding energy of the last neutron (or proton) is calculated, then certain values of N and Z have unusually high stability. These so-called magic numbers, which occur at N (and Z) = 20, 28, 50, 82, 126 correspond to closed nuclear shells and therefore are analogous to atomic closed shells. Closed atomic shells are caused by the strong central force of the nucleus. For this reason, the discovery of closed nuclear shells was surprising in view of the fact that strong central forces are not present in nuclei. This apparent dilemma was resolved by the shell model of the nucleus. It was pointed out that, although nucleons interact strongly, the exclusion principle makes it possible for nucleons to move in orbit inside a nucleus without making many collisions, and therefore individual nucleons move in the effective potential of the other nucleons. The shell model explains the location of the magic numbers by assuming that nucleons are subject to strong spin-orbit forces.

Silicon Burning

In Chapter 2 we discussed nucleosynthesis through carbon and oxygen burning. It might be assumed that at the end of these burning stages, nucleosynthesis would continue as a result of nuclear interactions between the products of carbon and oxygen burning such as Si^{28}, S^{32}, and Mg^{24}. However, the relatively high Coulomb barriers inhibit these reactions, and consequently photodisintegration of nuclei will occur after oxygen burning (that is, at temperatures in excess of $\simeq 3 \times 10^9 °K$). The principle of detailed balance allows us to express λ_{AB} the photodisintegration rate of nucleus AB in terms of the corresponding reaction rate between nuclei A and B. Under conditions of thermodynamic equilibrium, we have

$$\lambda_{AB} n_{AB} = n_A n_B \langle \sigma v \rangle \tag{9-34}$$

where the reaction rate $\langle \sigma v \rangle$ is defined in Equation (2-79). The ratio $(n_A n_B)/n_{AB}$ is determined by means of the Saha equation [see Equation (2-10)].

A great profusion of nuclear reactions will take place simultaneously when the temperature exceeds $\sim 3 \times 10^{9\circ}$K. Si^{28} is the last isotope of comparable atomic number to undergo photodisintegration because it is the most tightly bound. At high temperatures a quasiequilibrium will be set up between α-particle capture and the corresponding photodisintegration reactions. For temperatures less than $\sim 5 \times 10^{9\circ}$K, this quasiequilibrium, which is a consequence of the relative stability of Si^{28}, is described approximately by the set of reactions:

$$Si^{28} + He^4 \rightleftarrows S^{32} + \gamma$$
$$S^{32} + He^4 \rightleftarrows A^{36} + \gamma$$
$$A^{36} + He^4 \rightleftarrows Ca^{40} + \gamma$$
$$Ca^{40} + He^4 \rightleftarrows Sc^{44} + \gamma \qquad (9\text{--}35)$$
$$Sc^{44} + He^4 \rightleftarrows Ti^{48} + \gamma$$
$$Ti^{48} + He^4 \rightleftarrows Cr^{52} + \gamma$$
$$Cr^{52} + He^4 \rightleftarrows Ni^{56} + \gamma$$

where it has been assumed that the total number of protons (bound and free) is equal to the corresponding number of neutrons. The existence of a quasiequilibrium depends on the circumstance that S^{32}, A^{36}, and so on, undergo photodisintegration much more rapidly than Si^{28}. For this reason, the capture of α particles and the buildup of S^{32} are controlled by the inverse reactions.

The above series of reactions is called silicon burning, because the rate at which they will lead to the formation of α-particle nuclei is governed by the rate at which Si^{28} can be decomposed. Significant release of nuclear energy will result from silicon burning if the temperature is $< 5 \times 10^{9\circ}$K. For temperatures $\gtrsim 5 \times 10^{9\circ}$K, the very intense photon flux maintains an appreciable number of free nucleons and α particles. This circumstance leads to an appreciable reduction in the total binding energy and the final product of the reactions, which are not endothermic, is $Fe^{54} + 2p$. The quasiequilibrium that is set up during silicon burning is of finite duration because the decay of β radioactive nuclei leads to a reduction in the ratio Z/N which is initially unity. β decays will be important if the duration of silicon burning exceeds $\sim 10^4$ sec. If silicon burning takes place slowly (that is, at relatively low temperatures, $T < 3 \times 10^{9\circ}$K), β decays will be significant during silicon burning, and Fe^{56} can be produced directly. On the other hand, if silicon burning occurs at somewhat higher temperatures ($3 \times 10^9 \leq T \leq 5 \times 10^{9\circ}$K), silicon burning can be summarized by

$$2Si^{28} \to Ni^{56} \qquad (9\text{--}36)$$

which is exothermic by 10.9 MeV. Ni^{56} will subsequently β-decay to Fe^{56}.

In the temperature range $5 \times 10^9 < T < 10^{10}°K$, the quasiequilibrium will shift to Fe^{54}, and we have

$$2Si^{28} \rightarrow Fe^{54} + 2p \tag{9-37}$$

which is endothermic by -1.3 MeV. Because the solar abundance ratio Fe^{54}/Fe^{56} is low, the latter mode of silicon burning is probably not common in stars.

Recent calculations indicate that the simultaneous and explosive burning of carbon and silicon is a plausible way of explaining the relative solar abundances of common nuclei with atomic masses intermediate between carbon and nickel. Moreover, if such explosive burning is common in supernovae, then the production of gamma rays by means of the reactions

$$\begin{aligned}
Ni^{56} \ (e^-, \ \nu) \ Co^{56*} \\
Co^{56*} \rightarrow Co^{56} + \gamma \\
Co^{56} \ (e^-, \ \nu) \ Fe^{56} \\
Fe^{56*} \rightarrow Fe^{56} + \gamma
\end{aligned} \tag{9-38}$$

might produce a detectable flux of gamma rays.

e Process

If the rates of nuclear reactions and their inverses are strictly equal, nuclear statistical equilibrium will exist. The time scale over which equilibrium can be established is much shorter than the relevant β-decay time scales. For this reason, it is physically self-consistent to assume statistical equilibrium and at the same time ignore β decays. We note that β decays are not expected to participate in the statistical equilibrium since the inverse processes require absorption of neutrinos which will not occur at normal stellar densities. An example of such a process is

$$p + e^- \rightarrow n + \nu. \tag{9-39}$$

Under conditions of nuclear statistical equilibrium, the abundances of the various isotopes are determined by the density

$$\rho = \sum_{A,Z} Am_p N(A, Z) + m_p(n_p + n_n), \tag{9-40}$$

the temperature T, and the ratio

$$\frac{Z}{N} = \frac{\sum_{A,Z} ZN(A, Z) + n_p}{\sum_{A,Z} (A - Z)N(A, Z) + n_n} \tag{9-41}$$

where $N(A, Z)$ are the number densities of the isotopes. n_p and n_n are the number densities of the free protons and neutrons, respectively. In Equation

(9–40) we have neglected the proton-neutron mass difference. Under equilibrium conditions, the Saha equation [see Equation (2–10)] implies

$$\frac{N(A, Z)}{N(A - 1, Z)n_n} = \frac{g(A, Z)A^{3/2} \exp{(Q_n/kT)}}{2g(A - 1, Z)(A - 1)^{3/2}\theta} \tag{9–42}$$

where

$$\theta = \left[\frac{2\pi kT}{h^2}\left(\frac{M_{A-1}m_n}{M_{A-1} + m_n}\right)\right]^{3/2}$$

and Q_n is the binding energy of a neutron in the nucleus (A, Z). A similar relation involving free protons can be written

$$\frac{N(A - 1, Z)}{N(A - 2, Z - 1)n_p} = \frac{g(A - 1, Z)(A - 1)^{3/2} \exp{(Q_p/kT)}}{2g(A - 2, Z - 1)(A - 2)^{3/2}\theta} \tag{9–43}$$

where Q_p is the binding energy of the proton in the nucleus $(A - 1, Z)$. Successive application of the nuclear Saha equation gives

$$N(A, Z) = g(A, Z)A^{3/2}n_p^Z n_n^{A-Z}\theta^{1-A} \exp\left[\frac{Q(A, Z)}{kT}\right] \tag{9–44}$$

where

$$Q(A, Z) = c^2[Zm_p + (A - Z)m_n - M(A, Z)].$$

Equations (9–40), (9–41), and (9–44) show that the abundances $N(A, Z)$ are implicit functions of ρ, T, and Z/N.

The relative abundances of solar iron peak elements can be explained if it is assumed that matter has been brought into statistical equilibrium at $T \sim 5 \times 10^9{}°K$ and $\rho \sim 10^6$ g/cm³ with $Z/N \sim 0.85$. However, at high densities such as $\rho \sim 10^9$ g/cm³ and $T \sim 3 \times 10^9{}°K$, the abundances peak near Fe^{58} instead of Fe^{56}. Since the solar system Fe^{58}/Fe^{56} ratio is $<1/300$, it has been argued that e-process nucleosynthesis does not frequently occur at high densities.

r Process

Since the binding energy per nucleon decreases for isotopes heavier than the iron peak, nuclides heavier than $A \gtrsim 60$ are not expected to be produced during the nuclear burning stages of a star. Moreover, they are unlikely to be formed in stars by charged particle reactions because their high Z values imply that excessively high temperatures would be required for barrier penetration to be probable.

Neutron capture on already existing iron peak elements is the most plausible mechanism for producing heavy nuclides. In the s process, which we discussed in Section 5–4, neutron capture occurs on a 10^2–10^6 yr time scale, and consequently all energetically possible β decays are completed before additional neutron capture. Therefore, *s*-process nucleosynthesis produces nuclides that are close to the valley of β stability. The *s* process is terminated at Bi and Po, because the subsequent isotopes produced are short half-life α emitters. This circumstance implies that *s*-process nucleosynthesis cannot be responsible for the formation of U, Th, or a number of other stable, neutron-rich isotopes.

It has been suggested that rapid neutron capture (called *r* process) is responsible for the formation of U, Th, and other heavy neutron-rich isotopes. Neutron capture must take place in a time scale τ of about 10^{-5} sec per capture to sustain *r*-process nucleosynthesis. Supernova explosions are the likely sites for the very high neutron fluxes that are required. The neutron capture rate is

$$\frac{1}{\tau} = n\langle\sigma v\rangle \simeq n\sigma\bar{v} \tag{9-45}$$

where n is the neutron number density; $\sigma \sim 10^{-25}$ cm^2 is a typical cross section for neutron capture; and $\bar{v} \sim 3 \times 10^8$ cm/sec for an assumed temperature of $T \sim 3 \times 10^{9}$°K. It follows from Equation (9-45) that very high neutron densities ($n \simeq 10^{23}$ cm^{-3}) are necessary to produce neutron capture time scales as short as 10^{-5} sec. It is clear that very extreme physical conditions are required to produce such a high density of neutrons.

β decay can be neglected during *r*-process nucleosynthesis because the β-decay lifetimes are typically 10^{-2} sec. This circumstance implies that very neutron-rich isotopes are produced. However, after a large number of neutrons (~ 20) have been added to an isotope, the rate of neutron decay becomes as rapid as the rate of neutron capture, and consequently the nuclides remain until β decay can occur. After β decay, further neutron capture can produce still heavier isotopes. In this manner *r*-process nucleo-synthesis proceeds along a path in the *N-Z* plane that is approximately 15–30 neutrons in excess of the valley of β stability. Nuclides with $A \simeq 250$ are produced in about 5–10 sec. The *r* process terminates after isotopes somewhat more massive ($A \sim 275$) have been produced, because the spon-taneous fission half-life, which decreases with increasing nuclear charge, becomes shorter than the time scale for neutron capture. Fission cycles nuclides back into the chain of (n, γ) and β reactions. This cycling is pre-sumably stopped by the expansion that is caused by the supernova explosion. The nuclides that remain after the *r* process has ceased β decay back to stability in a predictable manner. Since nuclides that are close to magic numbers are more likely to remain at the termination of the *r* process than nuclides between magic numbers, *r*-process abundance peaks can be pre-dicted. The observed abundance peaks are in agreement with the predictions. This circumstance provides indirect evidence for the occurrence of *r*-process nucleosynthesis.

CHAPTER 10
Pulsars

10–1 INTRODUCTION

Pulsars are astronomical objects that emit radio pulses whose synchronization is so precise that their long-term stability approaches that of the most accurate atomic clocks. At least one pulsar, the Crab pulsar, has been discovered to emit x-ray and optical pulses as well as radio pulses. Figure 10–1 shows some examples of radio and x-ray pulses emitted by the Crab pulsar. Of the more than 40 known pulsars, the Crab pulsar has the shortest periodicity (0.033 sec). The pulsar with one of the longest presently known periodicities (2.7 sec) is also in the vicinity of the Crab nebula. A histogram showing the observed period distribution of pulsars is given in Figure 10–2. Measurements of the change in the periods of pulsars indicate that their periods are lengthening at the rate of $\Delta P/P \sim 10^{-13}\text{--}10^{-16}$.

The remarkable discovery of pulsars occurred by accident during a radio investigation of quasars. A careful examination of weak and spasmodic radio signals coming from a fixed point on the celestial sphere revealed that the signals were a succession of regularly spaced pulses with unpredictable amplitude. Pulsars were not discovered earlier because their emission is strongest at long wavelengths ($\lambda > 1$ m) rather than in the 1 m–3 cm wavelength range where most radio observations are made and also because the integration times used in most radio observations are much greater than pulsar pulse widths.

10–2 DISPERSION MEASURE AND THE INTERSTELLAR ELECTRON DENSITY

Signals of continuously changing wavelength characterize each pulse. A particular pulse is observed to arrive at a receiver tuned to a high frequency before it arrives at a receiver tuned to a lower frequency. The duration of

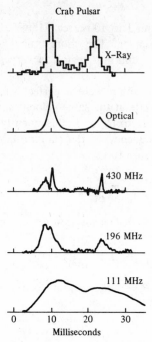

Figure 10–1   Mean pulse envelope of the crab pulsar at radio, optical, and x-ray wavelengths.

Source: From Hewish, A. 1970, *Ann. Rev. Astron. Ap.*, **8,** 265, by permission of Annual Reviews, Inc.

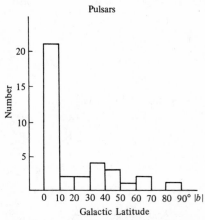

Figure 10–2   The distribution of pulsars in galactic latitude.

Source: From Hewish, A. 1970, *Ann. Rev. Astron. Ap.*, **8,** 265, by permission of Annual Reviews, Inc.

191

single pulses ranges from 1 to 100 msec. However, substructure as short as 100 μsec has been observed. The short time scales that characterize the pulses indicate that the emitting region is very small, probably less than 10^7 cm.

If the interstellar medium were a perfect vacuum, all electromagnetic radiation would propagate at the same velocity c. However, the interstellar medium contains free electrons, and consequently the propagation velocity will depend on the frequency of the radiation. An electromagnetic wave $E = E_0 e^{i(\omega t - kx)}$ will exert a force

$$m \frac{d^2 r}{dt^2} = -eE_0 e^{i\omega t} \tag{10-1}$$

on a free electron at $x = 0$. The solution for the displacement r is

$$r = \frac{e}{m\omega^2} E_0 e^{i\omega t}. \tag{10-2}$$

Since the polarization $P = -n_e er$, the dielectric constant of the medium is

$$\varepsilon = 1 + \frac{4\pi P}{E} = 1 - \frac{4\pi n_e e^2}{m\omega^2}. \tag{10-3}$$

The electric field vector E propagating in such a medium satisfies the wave equation

$$\nabla^2 \mathbf{E} = \frac{\varepsilon}{c^2} \frac{\partial^2 \mathbf{E}}{\partial t^2}. \tag{10-4}$$

Equations (10-3) and (10-4) yield the dispersion relation

$$\omega^2 = \omega_p^2 + k^2 c^2 \tag{10-5}$$

with

$$\omega_p = \left(\frac{4\pi e^2 n_e}{m}\right)^{1/2} = 5.6 \times 10^4 \sqrt{n_e}.$$

It follows that the propagation velocity is

$$\frac{\partial \omega}{\partial k} = \frac{kc^2}{\sqrt{k^2 c^2 + \omega_p^2}} \simeq \left(1 - \frac{1}{2} \frac{\omega_p^2}{\omega^2}\right) c \quad [\text{if} \quad \omega_p \ll \omega]. \tag{10-6}$$

The difference in propagation velocity between signals propagated at frequencies ω_1 and ω_2 is

$$\Delta\left(\frac{\partial \omega}{\partial k}\right) = \tfrac{1}{2} c \omega_p^2 \left(\frac{1}{\omega_2^2} - \frac{1}{\omega_1^2}\right)$$

$$\simeq \frac{c \omega_p^2}{\omega^3} \Delta\omega \quad \left[\text{if} \quad \frac{|\omega_1 - \omega_2|}{\omega_1} \ll 1\right]. \tag{10-7}$$

If D is the distance to the pulsar, then the differences in the arrival times of the two pulses is

$$\Delta t = \frac{D \, \Delta(\partial\omega/\partial k)}{c^2} . \tag{10-8}$$

From Equations (10–7), (10–8), and the definition of ω_p, we find

$$\Delta t = \frac{(3.1 \times 10^{+9} \int_0^D n_e \, dl) \, \Delta\omega}{c\omega^3} \tag{10-9}$$

where D is the distance between the pulsar and the observer. Equation (10–9) shows that the time delay of the pulse as measured at two frequencies ω_1 and ω_2 can be used to determine the dispersion measure $= \int n_e \, dl$. If the dispersion measure inside the source itself is small, then the interstellar electron density can be estimated if the distance is known, and conversely the distance to the pulsar can be estimated if the interstellar electron density is assumed known. Independent estimates of the mean distance to the observed pulsars ($\sim 10^3$ pc) allow a typical interstellar electron density of ~ 0.03 cm^{-3} to be determined.

10–3 ARGUMENTS FOR ROTATING NEUTRON STAR MODEL

It is widely believed that pulsars are rotating neutron stars. As we will show below, the observed range and remarkable constancy of their periods, as well as energy requirements that can be related to neutron star slowdown, provide convincing evidence in support of the rotating neutron star theory of pulsars. In addition, two pulsars, the Crab and Vela pulsars, are identified with supernova remnants.

The measured long-term constancy of pulsar periods shows that the pulses were not emitted from a planetary object or star in a binary system since the Doppler shift caused by orbital motion would shift their observed periodicity just as the earth's orbital motion does. If a source of period P is moving toward an observer with relative velocity v, the observed period is

$$P_1 = P \left(\frac{1 - (v/c)}{1 + (v/c)} \right)^{1/2} . \tag{10-10}$$

On the other hand, if the source is moving away from an observer, the observed period is

$$P_2 = P \left(\frac{1 + (v/c)}{1 - (v/c)} \right)^{1/2} . \tag{10-11}$$

From Equations (10–10) and (10–11) it follows that the change in period caused by orbital motion is

$$\frac{P_2 - P_1}{P} \equiv \frac{\Delta P}{P} = 2 \frac{v}{c} \quad \left(\text{if } \frac{\Delta P}{P} \ll 1 \right). \tag{10-12}$$

Since the measured values of $\Delta P/P$ are usually $\lesssim 10^{-15}$, the orbital velocity would have to be unreasonably low ($\lesssim 1.5 \times 10^{-5}$ cm/sec) if the emitting pulsars were part of a planetary or binary system. Since a substantial fraction of stars (especially massive stars) are found in binary systems, the absence of significant Doppler period shifts in pulsars argues that disruption of binary systems took place at the time pulsars were formed. This point was discussed in Section 9–2.

Because of the very high stability of the clock mechanism that regulates the time interval between pulses, the structure of the clock must remain very constant in time. For this reason, the clock mechanism must involve the pulsation or rotation of a star or the orbiting of two stars rather than periodicities associated with diffuse matter. Neutron stars and white dwarfs are the only types of stars that could maintain the observed stable periods and at the same time supply the necessary energy.

The existence of gravitational radiation rules out orbital motion as a plausible clock mechanism because the predicted change in angular frequency

$$\frac{d\omega}{dt} = -\frac{32G}{5c^5 I\omega} \left(\frac{M_1 M_2}{M_1 + M_2}\right)^2 D^4 \omega^6 \tag{10–13}$$

is much faster than the observed change. D is the distance between stars of mass M_1 and M_2 and I is the moment of inertia.

Although centrifugal forces help stabilize pulsation (that is, shorten the fundamental pulsation period), it is unlikely that rotation could reduce the periods of white dwarfs sufficiently ($\lesssim 0.1$ sec) to explain the periodicities of some pulsars. Although the high-order modes of white dwarfs have periodicities that are sufficiently short to explain pulsar observations, it is unlikely that these modes could maintain the stability necessary to explain pulsar observations. Moreover, no simple explanation for the observed slowdown of pulsars would follow. A rapidly rotating white dwarf is another possible model for a pulsar. However, the possible rotation periods of stable white dwarfs are not sufficiently short (except near the poles) to explain the shortest period pulsars.

Neutron stars are much more compact than white dwarfs, and consequently their fundamental pulsation periods and probable rotation periods are much less. The fundamental pulsation period of a self-gravitating configuration, which is approximately its free-fall time ($\sim (1/\sqrt{G\rho})$), will be less than 1 msec unless the mass of the neutron star is <0.1 M$\odot$. For this reason, and also because the damping of neutron star pulsations is expected to be much more rapid than indicated for pulsars, it is unlikely that pulsations are the pulsar clock mechanism.

By the process of elimination, the rotating neutron star model for pulsars is the most plausible. Moreover, if the observed slowdown of the Crab pulsar

$$\frac{1}{\nu}\frac{d\nu}{dt} = 10^{-13}$$

is related to the change in rotational energy of a neutron star, that is,

$$\frac{d(\frac{1}{2}I\omega^2)}{dt} = 4\pi^2 I\nu\frac{d\nu}{dt} \qquad (10\text{--}14)$$

with $\nu^{-1} = 0.033$ sec and $I \sim 10^{45}$ g-cm^2 for a typical neutron star, then the total power output is $\sim 2 \times 10^{38}$ ergs/sec, which is just the energy required to maintain the Crab nebula (see Section 9–5). This remarkable result is unlikely to be a coincidence.

10–4   SPATIAL DISTRIBUTION AND RATE OF PRODUCTION

The observed spatial distribution of pulsars shows that they are concentrated toward the galactic plane. Their distances lie in the range 10^2 pc $< d <$ 10^4 pc. Closer objects would be more isotropically distributed. On the other hand, more distant objects should be concentrated in the longitude quadrant toward the galactic center unless observational selection effects are important.

The above distance estimates allow us to arrive at a plausible lower limit on the total number of pulsars in our galaxy. If pulsars are distributed in a homogeneous manner in the vicinity of the galactic plane, then it is clear that the observed fraction represents less than 1 percent of the total number even if pulsar emission were isotropic and therefore could be seen by observers in all directions. Beaming of the radiation is likely to decrease the fraction of pulsars we observe by about a factor of 5. These arguments suggest that there are at least 10^5 pulsars in our galaxy. Since the measured slowdown rates of pulsars indicate that their observable lifetimes are $\sim 10^7$ yr, we arrive at the conclusion that approximately one pulsar is produced every 10^2 yr. Because this estimated rate of pulsar production is comparable to the estimated supernova production rate and also because neutron stars are likely remnants of supernova explosions, the discovery of pulsars provides evidence that a large number of supernova outbursts lead to the formation of neutron stars.

10–5   PERIOD DISTRIBUTION OF PULSARS

Figure 10–3 shows an apparent deficiency in the number of long-period pulsars. It has been suggested that the observed infrequency of long-period pulsars is related to the decay of magnetic fields in rotating neutron stars or to the tendency of the magnetic axis to align itself with the rotational axis

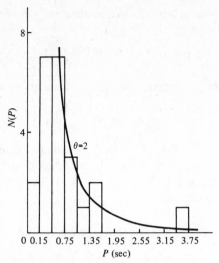

Figure 10–3   Observed period distribution of pulsars with 10° of the galactic plane compared with the predicted distribution for $\theta = 2$.

Source: From Setti, G. and L. Woltjer. 1970, *Ap. J.*, **159**, L87, by permission of the Chicago University Press.

after a sufficiently long period of time. It has been subsequently suggested, however, that neither of these explanations is required to explain the pulsar period distribution.

We consider a homogeneous distribution of pulsars and assume that the intrinsic period P and luminosity L are related to the age by the expressions

$$L = AP^{-\alpha}$$
$$P = Bt^{+\beta}. \qquad (10\text{–}15)$$

Let $N(L, >F)\, dL$ be the observed number of pulsars with absolute luminosity between L and $L + dL$ but with flux greater than $F = L/4\pi r^2$. Let $n(L)$ be the space density per unit luminosity interval of objects with luminosity L. We find

$$N(L, >F) = 4\pi \int n(L) r^2 \, dr$$

$$= \frac{1}{3} n(L) \left(\frac{L}{F}\right)^{3/2} \left(\frac{1}{4\pi}\right)^{1/2} \qquad (10\text{–}16)$$

where for simplicity it has been assumed that the pulsar space distribution is uniform and isotropic. The assumption of isotropy is not correct since pulsars tend to be concentrated toward the galactic plane. If pulsars are formed at a uniform rate R, we have

$$n(L) = R \frac{dt}{dL}. \qquad (10\text{-}17)$$

For $N(P, >F)$ we find

$$N(P, >F) = N(L, >F) \frac{dL}{dP}. \qquad (10\text{-}18)$$

Therefore, from Equations (10–15), (10–16), and (10–17), it follows that

$$N(P, >F) \propto P^{-\theta} \qquad (10\text{-}19)$$

with

$$\theta = -1 + \frac{1}{\beta} - \frac{3\alpha}{2}.$$

For the case of electromagnetic braking (see below), $L \propto P^{-4}$ and $P \propto t^{1/2}$. Therefore, $\theta = 5$ if the pulsed radio emission is a fixed fraction of the total energy release, and consequently very few long-period pulsars will be observed. Even for the case of $\theta = 2$, the number of observed long-period pulsars is sufficiently low to explain the observations.

Recent radio observations indicate that there is no simple relationship between pulsar radio emission and total energy release as inferred from pulsar slowdown. For this reason, the above explanation for the deficiency of long period (>3 seconds) pulsars is inadequate. A satisfactory argument must await a more complete understanding of physical conditions on the surfaces of rotating neutron stars.

10–6 LOSS OF ANGULAR MOMENTUM

If we accept the point of view that pulsars are rapidly rotating neutron stars, then measurements of the first and second time derivatives of their periods provide information about the physical processes responsible for their increasing periods. The rate of change of angular momentum J and rotational energy E are related by the expression

$$\frac{dJ}{dt} = \frac{d(I\omega)}{dt} = \frac{1}{\omega} \frac{dE}{dt} \qquad (10\text{-}20)$$

where ω is the angular frequency of rotation. We can write

$$\frac{dE}{dt} = K\omega^{n+1} \qquad (10\text{-}21)$$

where K is a factor independent of ω. From Equation (10–21) we find

$$\frac{d\omega}{dt} = \frac{K}{I}\,\omega^n$$

and

$$\frac{d^2\omega}{dt^2} = n\left(\frac{K}{I}\right)^2 \omega^{2n-1}.$$

(10–22)

Equations (10–22) imply

$$n = \frac{\omega(d^2\omega/dt^2)}{(d\omega/dt)^2}.$$

(10–23)

The measurable quantity n defined in Equation (10–23) can be related to the physical process or processes responsible for pulsar slowdown. A value $n = 5$ indicates quadrupole radiation (either gravitational or electric). A value $n = 3$ indicates magnetic dipole radiation while $n = 1$ indicates a stellar wind [see Equation (2–115) and discussion below]. If gravitational radiation can be neglected, the quantity n is a measure of the geometry (spherical or dipolar) of the magnetic field that surrounds the neutron star.

Neutron stars are likely end products of supernova explosions. During such outbursts the stellar core implodes, and the envelope is dispersed into space at very high velocities ($\sim 10^4$ km/sec). If the angular momentum of the imploding core is conserved or nearly conserved during collapse, then the frequency of rotation will increase as $\sim r^2$ where r is the radius of the collapsing core. The initial radius and central density of the core are likely to be $\sim 5 \times 10^8$ cm and $\sim 3 \times 10^9$ g/cm^3, respectively, while the final radius and density are $\sim 10^6$ cm and $\sim 10^{15}$ g/cm^3. This means that the period of rotation of the core will decrease by a factor of 10^6 during collapse. Therefore, its final rotation period may approach the neutron star limiting rotation period, which is approximately its fundamental pulsation period (~ 0.5 msec). The rotational energy of a rotating neutron star may exceed the energy necessary to explain the supernova light curve ($\sim 10^{50}$ ergs).

The presence of high magnetic fields in pulsars is consistent with the point of view that they are neutron stars. The high conductivity of the stellar interior ensures conservation of magnetic flux during collapse. For this reason, the magnetic field strength will increase as r^{-2} during collapse, and consequently magnetic field strengths as high as 10^{12} G are not unreasonable for neutron stars. Such high magnetic fields are required if the observed slowdown of pulsars is the result of electromagnetic radiation.

Gravitational Radiation

Gravitational radiation is the likely principal mechanism for the loss of angular momentum immediately after a supernova explosion. Deviations from axial symmetry are required for gravitational radiation to occur since the lowest-order radiation term is quadrupole radiation.[1] If the rotating neutron star configuration is assumed to be an ellipsoid

$$\frac{x^2}{\alpha^2} + \frac{y^2}{\beta^2} + \frac{z^2}{\gamma^2} = 1$$

with z the axis of rotation, then it can be shown that the rate of energy loss by means of gravitational radiation is

$$\frac{dE}{dt} = -\frac{32GI^2\varepsilon^4}{5c^5}\omega^6 \tag{10-24}$$

where the eccentricity is given by $\varepsilon = \sqrt{\alpha^2 - \beta^2}/\alpha$. It can be readily shown by means of Equation (10–24) that the eccentricity ε of the Crab pulsar would have to exceed 10^{-2} if gravitational radiation were to exceed the observed rate of energy production in the nebula. Very large magnetic fields ($B > 10^{15}$ G) would be required to produce this much distortion in a neutron star.

   If the slowdown of the Crab pulsar were caused primarily by gravitational radiation, then the energy half-life could be predicted from Equation (10–24). If we make the substitution $z = \omega^{-4}$, Equation (10–24) reduces to

$$\frac{dz}{dt} = -4K \tag{10-25}$$

where I is assumed constant, and $K = \frac{32}{5}GI(\varepsilon^4/c^5)$. Integrating Equation (10–25), we find

$$z = 4Kt + \text{constant}$$

or $\tag{10-26}$

$$\omega^{-4} = 4Kt + \omega_0^{-4}$$

where $\omega = \omega_0$ at $t = 0$. The rotation energy as a function of time becomes

$$E(t) = \tfrac{1}{2}I\omega^2 = \tfrac{1}{2}I(4Kt + \omega_0^{-4})^{-1/2}. \tag{10-27}$$

At $t = 0$ we have $E(0) = \tfrac{1}{2}I\omega_0^2$. Equations (10–27) shows that a neutron star would lose half its energy in a time

$$t_{1/2} = \frac{3}{4K\omega_0^4}. \tag{10-28}$$

The present angular frequency of the Crab pulsar is $\omega = 190$ rad/sec. The time interval for a neutron star with twice the energy of the present Crab pulsar to slow down to the present rotation rate is ~ 5000 yr, which is

---

[1] See Appendix D.

considerably longer than the known age of the Crab nebula (915 yr). More-over, the measured value of the quantity n defined in Equation (10–23) would be 5 if gravitational radiation or electric quadrupole radiation were responsible for the observed slowdown. This predicted value for n is much larger than the measured value $n \sim 2.6$.

Stellar Winds

Mass loss by means of stellar winds is another mechanism by which a neutron star can lose angular momentum. In the absence of a significant surface magnetic field (that is, if the Alfvén velocity, $v_A = \sqrt{B^2/4\pi\rho}$, is less than the usual sound velocity), the rate of change of angular momentum of a star with radius R that is losing mass at a rate dM/dt is

$$\text{Torque} = \frac{dJ}{dt} = -\frac{2}{3}\frac{dM}{dt}R^2\omega \tag{10–29}$$

where ω is the angular velocity of the star. When the surface magnetic field is appreciable, the magnetosphere will co-rotate with the star, and the torque on the star is proportional to ω [see Equation (2–115)]. Co-rotation, which is possible only so long as Alfvén waves can transmit shear upstream, cannot extend beyond the speed of light cylinder defined by $r_0 = c/\omega$, where r_0 is the perpendicular distance from the axis of rotation. For pulsars, the motions of the magnetosphere will be relativistic, and therefore the results given in Equation (2–115) are not strictly applicable. However, it can be shown that except for a factor of order unity, the expression for the torque given by Equation (2–115) is valid if v is replaced by c. If stellar winds were the principal mechanism for the loss of angular momentum, then the magnetic field would be primarily radial and the value of n would be unity, which is lower than indicated for the Crab pulsar.

Magnetic Dipole Radiation

A rotating configuration with a magnetic dipole field that is surrounded by a vacuum will radiate magnetic dipole radiation so long as the magnetic dipole vector **m** does not lie along the axis of rotation. If θ is the angle between the rotation axis z and the magnetic moment vector **m**, we have

$$\frac{d^2m_\perp}{dt^2} = m \sin \theta \omega^2$$

$$= \frac{Ba^3}{2} \sin \theta \omega^2 \tag{10–30}$$

where ω is the angular frequency of rotation; B is the magnetic field strength at the poles; and a is the radius of the configuration. Therefore, the energy radiated by a time-varying dipole moment

$$\frac{dE}{dt} = -\frac{2}{3c^3}\left(\frac{d^2 m_\perp}{dt^2}\right)^2 \tag{10-31}$$

becomes

$$\frac{d(\frac{1}{2}I\omega^2)}{dt} = \frac{dE}{dt} = -\frac{2m^2 \sin^2 \theta \omega^4}{3c^3}. \tag{10-32}$$

If we substitute $a = 10^6$ cm, $B = 10^{12}$ G, $\theta = \pi/4$, and $\omega = 10$ rad/sec into Equation (10–32), we find $dE/dt \sim 10^{32}$ ergs/sec, which is comparable to the energy required to explain the observed slowdown of most pulsars. The predicted value $n = 3$ for dipole radiation is much closer to the observed value of n than that predicted by quadrupole radiation.

The assumption that dipole radiation is responsible for the slowdown of the Crab pulsar leads to a predicted age that is close to the known historical age. If we substitute $z = \omega_0/\omega$ and

$$K = \frac{2m^2 \sin^2 \theta \omega_0{}^2}{3c^3 I}$$

Equation (10–32) becomes

$$\frac{dz^2}{dt} = 2K. \tag{10-33}$$

Solving Equation (10–33), we find

$$t = \frac{1}{2}\frac{z^2}{K} + \text{constant} \tag{10-34}$$

with $\omega = \omega_0$ at $t = 0$. If t is set equal to zero at the present time, then the age of the pulsar, which is defined to be the time at which its rotation period was very short, becomes

$$t_{\text{age}} = -\frac{1}{2K}. \tag{10-35}$$

If we substitute the observed variables for the Crab pulsar into Equation (10–32), we find $B \sim 2 \times 10^{12}$ G. Using this value for B in Equation (10–35), the calculated age of the Crab pulsar becomes close to the known lifetime of the nebula.

The magnetic dipole model neglects the effect of a plasma surrounding the neutron star. This model is plausible for neutron stars because the very intense magnetic fields may produce radiation pressure sufficient to sweep away the surrounding plasma. There is observational evidence that the electron densities are low near the center of the Crab nebula. Moreover, as we shall show below, even if there is a significant plasma surrounding a neutron star, the resultant radiation may be similar to that of a magnetic dipole rotating in a vacuum.

10–7 ROTATING NEUTRON STAR WITH MAGNETOSPHERE

A rotating neutron star with a strong external magnetic field is assumed to be surrounded by a plasma. The presence of a strong magnetic field in a plasma will make the electrical conductivity anisotropic. The conductivity will be relatively low in directions transverse to the magnetic field, but very high along magnetic lines of force. Since $J = \sigma E$ (generalized Ohm's law), it is reasonable to expect that if excessive current densities are to be available, the condition $E \cdot B = 0$ will be approximately satisfied in a magnetosphere surrounding a rotating neutron star.

Figure 10–4 shows a plausible magnetic field configuration for a pulsar. The magnetic field is assumed to be approximately that of a dipole far inside the speed of light cylinder ($\omega r \sin \theta \ll c$). However, as $\omega r \sin \theta \to c$, the field becomes increasingly distorted from dipolar form as a result of the motion of charged particles along the magnetic field lines. Charges that move along magnetic field lines that close within the light cylinder co-rotate with the star. The co-rotating magnetosphere is bounded by magnetic field lines that make an angle

$$\sin \theta_0 \sim \left(\frac{\omega R}{c}\right)^{1/2}$$

with respect to the axis of rotation (R is the radius of the star, and the magnetic field is assumed dipolar). That the angle θ_0 represents the approximate boundary of the magnetosphere can be seen from Figure 10–4 where it has been assumed that the magnetic field lines that bound the co-rotating magnetosphere are tangent to the light cylinder. Lines of force with $\theta < \theta_0$ are assumed to extend beyond the light cylinder.

The potential difference between θ_0 and the pole is

$$\Delta V = \frac{\omega B R^2}{2c} \sin^2 \theta_0 \sim \frac{1}{2}\left(\frac{\omega R}{c}\right)^2 RB. \tag{10–36}$$

Equation (10–36) indicates that the most energetic escaping particles should attain energies of $\frac{1}{2}Ze\,\Delta V$ by the time they reach large distances from the star. From Equation (10–36) the maximum energy becomes

$$E_{max} \sim 3 \times 10^{12} \frac{Z R_6{}^3 B_{12}}{P^2} \text{ eV} \tag{10–37}$$

where Z is the charge of the particle; R_6 is the radius of the star in units of

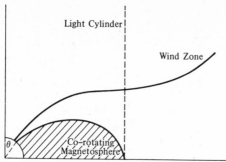

Figure 10-4 Rotating neutron star with magnetic field. Magnetic axis is chosen to coincide with the rotational axis of star. Magnetic lines of force that contact the rotating neutron star at $\theta < \theta_0$ extend beyond the light cylinder.

10^6 cm; B_{12} the magnetic field strength in units of 10^{12} G; and P the period of the pulsar in seconds.

It has been argued that the pulsar radiation field exists beyond the light cylinder. If the nonzero electromagnetic field components are assumed to be of the form

$$B_r = \frac{\psi(\theta)}{r^2} \qquad (10\text{-}38)$$

and

$$B_\phi = E_\theta = -\frac{\omega r}{c}\sin\theta\, B_r$$

then the total amount of radiated energy can be obtained by integrating the radial component of the Poynting vector

$$\mathbf{S} = \frac{c}{4\pi}(\mathbf{E} \times \mathbf{B}) \qquad (10\text{-}39)$$

over the surface of a sphere centered at the center of the star. The radial component of the Poynting vector is

$$S_r = \frac{c}{4\pi}E_\theta B_\phi = \frac{\omega^2}{4\pi c}\sin^2\theta\,\frac{\psi^2(\theta)}{r^2}. \qquad (10\text{-}40)$$

Equation (10-40) implies

$$\frac{dE}{dt} = 2\int_0^{2\pi}\int_0^{\theta_0} S_r r^2 \sin\theta\, d\theta\, d\phi$$

$$= \frac{\omega^2}{c}\int_0^{\theta_0}\psi^2(\theta)\sin^3\theta\, d\theta \equiv \frac{\omega^2}{c}I. \qquad (10\text{-}41)$$

Recalling that

$$\Phi = 2\pi \int_0^{\theta_0} \psi(\theta) \sin\theta \, d\theta$$

$$\sim B\pi R^2 \theta_0{}^2 \sim \frac{1}{2} \frac{\omega R}{c} R^2 B \tag{10-42}$$

is the magnetic flux leaving the star, plausible choices of the function $\psi(\theta)$ give $I/\Phi^2 \sim 1$ and therefore the torque on the star is

$$\frac{dJ}{dt} = \frac{1}{\omega} \frac{dE}{dt} \sim R^3 B^2 \left(\frac{\omega R}{c}\right)^3 \tag{10-43}$$

where R is the radius of the star, and B is the magnetic field at the poles. Equation (10-43) predicts a rate of slowdown that is similar to that predicted by means of magnetic dipole radiation [see Equation (10-32)].

10-8   PULSARS AND THE ORIGIN OF COSMIC RAYS

The origin of cosmic-ray particles (especially very-high-energy particles) has remained a mystery for decades. Although the discovery of pulsars has not solved the problem of the origin of cosmic radiation, it has provided important new clues about how high-energy particles are accelerated. Observations of the Crab nebula and pulsar make it clear that the Crab pulsar is producing high-energy electrons at a rate that is comparable with the loss of its rotational energy. The measured slowdown of pulsars (most notably the Crab pulsar) shows that the rotational energy of a compact object can be efficiently transformed into high-energy particles.

If cosmic-ray particles are of galactic origin and have lifetimes that are $\sim 10^7$ yr, then the rate of cosmic-ray production necessary to maintain the galactic cosmic-ray energy density in a steady state is $\sim 10^{40}$ ergs/sec for protons and heavy ions and $\sim 4 \times 10^{38}$ ergs/sec for cosmic electrons. Supernova outbursts have sufficient energy to supply the necessary production rate. Moreover, radio emission from supernova shells and the existence of pulsars provide strong evidence that high-energy particle production is associated with supernovae. However, at the present time it is not clear that supernova outbursts are the primary source of cosmic radiation. It appears that unless cosmic-ray production is very high just after a supernova outburst, pulsars do not supply enough high-energy particles to maintain the measured cosmic-ray flux.

Even before the discovery of pulsars, it had been recognized that cosmic-ray particles can be accelerated in supernova shells. However, the highest-energy particle of charge Z that can be produced in a region of characteristic length l and magnetic field strength B is [see also Equation (10-37)]

$$E_{\max} \sim eZBl = 300ZBl \text{ eV.} \tag{10-44}$$

If we use the physical parameters associated with the Crab nebula, which is the most favorable known supernova remnant (that is, $B \sim 10^{-4}$ G, $l \sim 10^{18}$ cm), we find $E_{max} \sim 3 \times 10^{16} Z$ eV. Although this is an impressively high energy for a single particle, it is not sufficiently high to explain the highest-energy cosmic-ray particles (10^{20}–10^{21} eV). On the other hand, a pulsar with $B \sim 10^{12}$ G and $l \sim 10^6$ cm could, in principle, accelerate particles to $E_{max} \sim 3 \times 10^{20} Z$ eV, which is sufficient to explain the highest-energy particles presently detected. Thus, rapidly rotating neutron stars represent a plausible galactic site for the production of high-energy particles. Measurements of the isotropy of very-high-energy particles are needed to establish whether or not they are produced within our galaxy.

10–9 PULSAR RADIATION

Although the total pulsed radiation from pulsars appears to be much less than the total energy loss, the origin of pulsed radiation is of considerable interest since an understanding of its nature would provide important information about the physical conditions inside pulsar magnetospheres. Pulsar radiation is likely to arise from high-energy particles (electrons, protons, or heavy ions) constrained to stream along magnetic lines of force that begin near magnetic poles and extend beyond the speed of light cylinder. The existence of a beam of high-energy particles with small pitch angle is physically plausible in a plasma containing a high magnetic field, because particles that do not have small pitch angles will very rapidly lose their energy by means of synchrotron radiation (see Section 9–3).

A plausible but very preliminary model for pulsar radiation assumes that the radio emission is the result of acceleration caused by constraining particles to move along curved magnetic lines of force. On the other hand, the optical and x-ray emission from the Crab pulsar is possibly caused by the simultaneous gyration of particles around these same magnetic lines of force. This theoretical point of view is supported by the observed simultaneity of x-ray and radio pulses.

From the theory of synchrotron radiation, the power per unit frequency interval emitted by a relativistic ($\gamma = E/mc^2 \gg 1$) particle moving along an orbit with radius of curvature r_c is

$$P(v) = \frac{1}{2\pi} \frac{e^2}{c} \left(\frac{c}{r_c}\right) \gamma \left(\frac{v}{v_c}\right)^{1/2} \text{ erg-sec}^{-1}\text{-Hz}^{-1} \qquad \text{(for } v < v_c). \qquad (10\text{–}45)$$

The maximum value of $P(v)$ arises just below the critical frequency

$$v_c \sim \frac{1}{2\pi} \frac{c}{r_c} \gamma^3. \qquad (10\text{–}46)$$

The observed radio brightness temperature T_b from the Crab pulsar is $\sim 10^{26}$°K if the characteristic dimension of the emitting region is 10^7 cm. Such high brightness temperatures are possible if the radiation is coherent (that is, the particles move in bunches smaller than a wavelength). Let us estimate some hypothetical properties of an electron beam that might explain the observed properties of pulsed radiation from the Crab pulsar. For the present let us assume that the pulsed radiation is produced close to the speed of light cylinder, which is defined by the distance

$$r \sim \frac{c}{\omega} \sim 1.5 \times 10^8 \text{ cm} \tag{10-47}$$

from the center of the Crab pulsar. If the magnetic field is assumed to be $\sim 2 \times 10^{12}$ G at the poles of the pulsar and its radius is assumed to be 10^6 cm, the magnetic field will be on the order of 10^6 G in the vicinity of the speed of light cylinder. The observed timing of the pulses indicates that the radiating particles are moving at small pitch angles, that is,

$$\theta \lesssim \frac{\tau}{P} \tag{10-48}$$

where τ is the width of the pulse, and P is the period of the pulsar. For the Crab pulsar we have $\tau \sim 250$ μsec and $P \sim 33$ msec which imply $\theta \leq 8 \times 10^{-2}$. Therefore, the perpendicular component of the magnetic field becomes $B_\perp \leq 8 \times 10^4$ G.

The following arguments allow us to estimate the number of particles in a bunch. Since coherent radiation is observed at wavelengths as short as 30 cm, Equation (10–46) indicates that γ exceeds 300. The required number of particles in a bunch must satisfy the relation

$$N \geq \frac{kT_b}{mc^2\gamma}. \tag{10-49}$$

Thus N is $\sim 10^{14}$ for the Crab pulsar since γ is about 300 and $T_b \sim 10^{26}$°K. The required number density of electrons n_e can be estimated from the approximate expression for coherent synchrotron radiation, that is,

$$\frac{1}{2\pi} \frac{e^2}{c} \left(\frac{c}{r_c}\right)^2 \gamma^4 n_e V N \sim 10^{31} \text{ ergs-sec}^{-1} \tag{10-50}$$

where 10^{31} ergs/sec is the radio luminosity of the coherent pulsed radiation, which is emitted over a bandwidth $\sim 10^9$ Hz. Substituting $N \sim 10^{14}$, $c/r_c \sim 200$ sec^{-1} and $V \sim l^3 \sim 10^{21}$ cm^3 in Equation (10–50), we find $n_e \sim 10^{12}$–10^{13} cm^{-3}.

In the above discussion we have assumed that the pulsed emission arises from relativistic electrons streaming in the vicinity of the light cylinder. The observed high degree of linear polarization and rotation of the plane of polarization within a single pulse width has led to the suggestion that the radiation arises well inside the light cylinder and therefore at higher values of magnetic field. If the pulsed radiation were emitted close to the star, synchrotron self-absorption might be important for relativistic electrons. This circumstance has led to the suggestion that radiation from relativistic protons close to the magnetic poles produces the pulsed radiation.

Equation (10–46) shows that if magnetic constraint radiation were responsible for the pulsed x-ray emission ($v \sim 10^{18}$ Hz) from the Crab pulsar, a value of $\gamma \sim 10^5$ would be required for the relativistic particles. Since the x-ray luminosity is $\sim 10^{36}$ ergs/sec and the bandwidth $\sim 10^{18}$ Hz, the number of relativistic particles in the emitting region would follow from the expression

$$\frac{1}{2\pi} \frac{e^2}{c} \left(\frac{c}{r_c}\right) \gamma \cdot n_e V \sim 10^{18} \text{ ergs-sec}^{-1}\text{-Hz}^{-1}. \tag{10-51}$$

Equation (10–51) leads to a predicted number density of relativistic particles in the emitting region that is excessively large ($\sim 10^{40}$–10^{41}).

It is plausible that the optical and x-ray emission from the Crab pulsar is caused by normal incoherent synchrotron radiation whose critical frequency is

$$v_c \sim \frac{eB\gamma^2}{2\pi mc}. \tag{10-52}$$

It has been suggested that the scattering of particles with small pitch angles by means of plasma particle-wave interaction could lead to the observed optical and x-ray pulsed radiation. As the pulsar slows down, the light cylinder will move to greater distances from the axis of rotation. Consequently, if the incoherent x-ray and optical radiation arises from the vicinity of the light cylinder, the observed intensity of the radiation should decrease rapidly. This point of view regarding the origin of pulsar radiation provides a plausible explanation for why only the most rapidly rotating pulsar, namely the Crab pulsar, emits detectable optical or x-ray radiation.

CHAPTER 11
Neutron Stars

11-1 INTRODUCTION

White dwarfs (more correctly black dwarfs) undoubtedly represent the final equilibrium states of many stars. However, no equilibrium configuration can be supported by means of electron degenerate matter if its mass exceeds the Chandrasekhar mass limit or if it is somehow compacted to higher than white dwarf densities. The possible existence of equilibrium configurations of very high density is of fundamental importance to both astronomy and physics. As will be indicated below, such very dense configurations, which are called neutron stars, are predicted on the basis of theoretical arguments. Moreover, the recent discovery of pulsars provides convincing evidence that such objects are common in nature. Neutron stars are unique among stable stars, because their basic properties are influenced by general relativistic effects and also because nuclear forces play a dominant role in determining their equation of state. It should be emphasized that, although the term "neutron star" is commonly used as a generic term for all very dense equilibrium configurations, the number densities of mesons and hyperons will exceed that of neutrons at very high densities (that is, $\rho \gtrsim 10^{15}$ g/cm³).

11-2 TOLMAN-OPPENHEIMER-VOLKOFF EQUATION

The space-time interval ds is defined by the expression

$$ds^2 = g_{ij} \, dx^i \, dx^j \qquad (i = 1, 2, 3, 4) \qquad (11\text{--}1)$$

where the dx^i's are coordinate distances, and the g_{ij} is a tensor called the metric tensor. Since ds^2 is a scalar, it is invariant under coordinate transformations. In the absence of a gravitational field, ds^2 can be written

$$ds^2 = c^2 \, dt^2 - (dx^2 + dy^2 + dz^2) \qquad (11\text{--}2)$$

which is the same in all inertial (that is, Lorentzian) frames of reference. For a static, spherically symmetric mass, a coordinate system can be found such that

$$ds^2 = e^{v(r)} dt^2 - e^{-\lambda(r)} dr^2 - r^2 (d\theta^2 + \sin^2 \theta \, d\phi^2). \quad (11\text{--}3)$$

The general theory of relativity is based on the principle of equivalence which asserts that at every space-time point, we can choose a coordinate system such that in the neighborhood of this point, the laws of physics have the same form as they do in the absence of a gravitational field. The measured equality of the inertial and gravitational mass provides the experimental basis for the principle of equivalence.

The Einstein field equations, which relate the curvature of space and the distribution of matter (and other forms of energy) throughout space, can be written

$$R_{ij} - \tfrac{1}{2}g_{ij}R = -\frac{8\pi}{c^4} G T_{ij} \quad (11\text{--}4)$$

where the left-hand side of Equation (11–4) is a tensor that depends on g_{ij} and its first and second derivatives, T_{ij} is the energy-momentum tensor. In a frame of reference that is locally Lorentzian, the energy-momentum tensor for an isotropic fluid is

$$T_{ij} = \begin{pmatrix} \rho c^2 & 0 & 0 & 0 \\ 0 & P & 0 & 0 \\ 0 & 0 & P & 0 \\ 0 & 0 & 0 & P \end{pmatrix} \quad (11\text{--}5)$$

where P is the pressure, and ρc^2 is the total energy density. When used in conjunction with Equations (11–3) and (11–5), the Einstein field equations lead directly to three independent equations:

$$e^{-\lambda(r)} \left[\frac{1}{r} \frac{d\lambda(r)}{dr} - \frac{1}{r^2} \right] + \frac{1}{r^2} = \frac{8\pi G \rho}{c^4} \quad (11\text{--}6)$$

$$\frac{dP}{dr} + \frac{1}{2} \frac{dv(r)}{dr} \left[\rho c^2 + P \right] = 0 \quad (11\text{--}7)$$

$$e^{-\lambda(r)} \left[\frac{dv(r)}{dr} \frac{1}{r} + \frac{1}{r^2} \right] - \frac{1}{r^2} = \frac{8\pi G}{c^2} P. \quad (11\text{--}8)$$

Equation (11–6) can be rewritten in the form

$$\frac{d}{dr} (re^{-\lambda(r)}) = 1 - \frac{8\pi G}{c^2} r^2 \rho. \quad (11\text{--}9)$$

Integrating, we find

$$e^{-\lambda(r)} = 1 - \frac{8\pi G}{rc^2} \int_0^r \rho r^2 \, dr. \quad (11\text{--}10)$$

In the present discussion, the quantity ρc^2 includes not only the rest mass energy, but also the kinetic energy and interaction energy (excluding gravity). At the surface of the star, the line element given by Equation (11–3) must reduce to the corresponding expression in a vacuum that surrounds a mass M, that is,

$$ds^2 = -\left(1 - \frac{2GM}{c^2 r}\right)^{-1} dr^2 - r^2 (d\theta^2 + \sin^2 \theta \, d\phi^2)$$

$$+ \left(1 - \frac{2GM}{c^2 r}\right) c^2 \, dt^2 \tag{11–11}$$

where M is the total rest mass energy.

From the equations above, we can identify the mass-energy interior to r as

$$M_r = 4\pi \int_0^r \rho r^2 \, dr. \tag{11–12}$$

The expression for mass-energy given in Equation (11–12) includes both the positive work of compression and the negative potential energy due to gravitation.

If we eliminate $(dv(r)/dr)$ from Equations (11–7) and (11–8), we find

$$\frac{dP}{dr} = \frac{-e^{\lambda(r)}[\rho c^2 + P][-e^{-\lambda(r)} + r^2(8\pi G/c^4)P + 1]}{2r}. \tag{11–13}$$

Substituting the expression for $e^{-\lambda(r)}$ from Equation (11–10) into Equation (11–13), we find

$$\frac{dP}{dr} = -\frac{GM_r(\rho + (P/c^2))(1 + (4\pi P r^3/M_r c^2))}{r^2(1 - (2GM_r/rc^2))} \tag{11–14}$$

which is the TOV equation. This equation describes hydrostatic equilibrium in general relativity.

If we compare the above equation to the corresponding Newtonian equation, we see that the effect of general relativity is due to the curvature of space and pressure regeneration. The curvature of space becomes important as

$$r \to \frac{2M_r G}{c^2} \equiv r_s \tag{11–15}$$

where r_s is the Schwarzschild radius. Pressure regeneration arises because, as we compress matter, the resultant increase in pressure increases the effective gravity and therefore, as shown in Equation (11–14), the pressure gradient is increased over what would be expected from Newtonian theory. The two terms in the TOV equation that reflect pressure regeneration are significant when

$$P \sim \rho c^2. \tag{11–16}$$

These terms generally become important at lower densities than the curvature of space term.

In general relativity, the term $e^{(v(r))/2}$ given in Equation (11–3) plays a role similar to that of potential energy in Newtonian theory. That this is the case can be seen from the following argument. Allow a narrow tube to extend from the surface of the neutron star into the interior. A photon that is emitted at a radius r with a frequency f in the local Lorentz frame will be redshifted to a frequency f' at infinity where

$$f' = fe^{-(v(r)/2)} \tag{11–17}$$

or

$$\frac{\Delta f}{f} = \frac{f' - f}{f} = 1 - e^{-(v(r)/2)}.$$

In a weak gravitational field, Equation (11–17) reduces to

$$\frac{\Delta f}{f} = - \frac{GM_r}{rc^2}. \tag{11–18}$$

Equations (11–17) and (11–18) make the correspondence between $e^{(v(r))/2}$ in general relativity and the Newtonian potential energy apparent.

11–3   EQUATION OF STATE

Equations (11–12) and (11–14) are sufficient to determine the structure of a neutron star if an equation of state of the form

$$P = P(\rho) \tag{11–19}$$

is specified. The expected equation of state of a neutron star can be conveniently divided into several domains. From low densities up to densities of about 10^{10} g/cm^3, the equation of state is generally assumed to be that of a degenerate electron gas. For densities in the range $10^{10} \lesssim \rho \lesssim 4 \times 10^{12}$ g/cm^3, one assumes that heavy nuclei are in beta equilibrium with a relativistic degenerate electron gas.

There exists a line in the (Z, A) plane, which is known as the neutron drip line. If the neutron-proton ratio exceeds a certain limit, the nucleus is unstable to the loss of a neutron. For densities $\gtrsim 4 \times 10^{12}$ g/cm^3, electron capture shifts the equilibrium from heavy stable nuclei to nuclei unstable to neutron drip. Consequently, for densities in the range $4 \times 10^{12} \leq \rho \leq 2 \times 10^{14}$ g/cm^3, the equilibrium shifts to free nucleons (mostly neutrons), although some nuclei remain up to densities as high as nuclear matter. The conditions that govern equilibrium are charge conservation, which implies

$$n_e = n_p \quad (\text{cm}^{-3}) \tag{11–20}$$

and the condition

$$E_{Fe} + E_{Fp} = E_{Fn} \tag{11–21}$$

where E_{Fe}, E_{Fp}, and E_{Fn} are the Fermi energies of the electrons, protons, and neutrons, respectively. Since the number densities, n_e and n_p, are proportional to the third power of the Fermi momentum [Equation (8–9)], it follows from Equation (11–20) that

$$p_{Fe} = p_{Fp}. \tag{11–22}$$

At densities such that hyperon and meson production can be neglected (that is, $\rho \lesssim 10^{15}$ g/cm^3), the equilibrium is determined by the reaction

$$n + p + e^- \rightarrow n + n + \nu_e.$$

Although the electrons are relativistic particles in the density range under consideration, the neutrons and protons are nonrelativistic particles. Therefore, Equation (11–21) implies

$$cp_{Fe} + \frac{p_{Fp}^2}{2m_p} + m_p c^2 \simeq m_n c^2 + \frac{p_{Fn}^2}{2m_n} + E_\nu. \tag{11–23}$$

Since $p_{Fp} = p_{Fe}$, $m_p c^2 \sim m_n c^2$, $p_{Fp} \ll m_p c$, and E_ν is small, Equation (11–23) reduces to

$$cp_{Fe} \simeq \frac{p_{Fn}^2}{2m_n}. \tag{11–24}$$

Equations (8–9) and (11–24) imply

$$\frac{n_e}{n_n} = \left(\frac{p_{Fe}}{p_{Fn}}\right)^3 = \left(\frac{p_{Fn}}{m_n c}\right)^3 \ll 1. \tag{11–25}$$

The relation given in Equation (11–25) shows that the number density of neutrons exceeds that of electrons so long as mesons and hyperons are not dominant. Hyperons and mesons will become abundant when

$$\frac{p_{Fn}^2}{2m_n} \gtrsim 140 \text{ MeV} \tag{11–26}$$

which is about the rest mass energy of π mesons. This inequality shows that neutrons will be the dominant particle in the interior of a neutron star if the density is in the range $10^{13} \lesssim \rho \le 10^{15}$ g/cm^3, which is near the normal neutron density of a heavy nucleus (1.5×10^{14} g/cm^3).

If neutrons are assumed to be a free fermion gas, it can be shown by means of numerical integrations of Equation (11–12) and (11–14) that the maximum mass for a stable neutron star is $\simeq 0.7$ M$\odot$. More massive configurations would presumably collapse to arbitrarily high densities. However, at densities equal to or greater than normal nuclear densities, the equation of state of a neutron star will depend critically on nuclear many-body interactions, and consequently the free fermion equation of state is obviously an inadequate approximation.

Figure 11–1 shows several equations of state for neutron matter. These equations of state are compared to that of a free fermion gas. Curves 1 and 2 were obtained by solving the Brueckner-Bethe-Goldstone equation of state of nuclear matter with two different nuclear potentials. It is convenient to express the equation of state in the polytropic form $P = Kn^{\gamma}$, where P is the pressure and n the particle density. If the neutron gas is nonrelativistic, the rest mass energy of the neutrons exceeds their Fermi energies, and consequently ρ is nearly proportional to n. For a nonrelativistic free fermion gas, $\gamma = \frac{5}{3}$. The corresponding values of γ are 1.5 and 1.6 for the two nuclear equations of state shown in Figure 11–1. At a fixed density, the computed pressures are lower for the equation of state predicted by means of nuclear matter calculations than for the free fermion case. Dimensional analysis [see Equation (8–17)] indicates that for a fixed central density, the mass of a neutron star is less than predicted by means of a free fermion gas. Numerical calculations that include nucleon-nucleon interactions show that the limiting mass of a neutron star composed primarily of neutrons (that is, whose central density is $\lesssim 5 \times 10^{14}$ g/cm^3) is ~ 0.3 M$\odot$, which is well below the

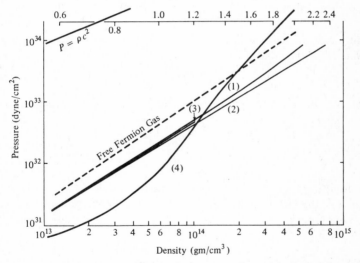

Figure 11–1   Equations of state for neutron matter. Curves 1, 2, and 3 are based on recent calculations with different nuclear potentials. Curve 4 is based on an earlier calculation. The dotted line corresponds to a noninteracting neutron gas. Curve $P = \rho c^2$ is the limiting pressure if the velocity of sound is to be less than the velocity of light.

Source: From Wang, C. G., W. K. Rose, and S. L. Schlenker. 1970, *Ap. J.*, **160**, 117, by permission of the Chicago University Press.

Chandrasekhar mass limit for white dwarfs ($M_c \sim 1.4$ M$\odot$) as well as the TOV result for a free fermion gas.

It should be pointed out that if the equation of state (that is, $P = P(\rho)$) were to stiffen at densities in excess of 5×10^{14} g/cm^3 as a result of a hard core component associated with the nuclear potential, then the limiting mass of a neutron star could exceed the limiting white dwarf mass. Nuclear matter calculations are exceedingly complicated and not correct for densities greater than about 5×10^{14} g/cm^3. It is, therefore, possible that the actual nuclear equation of state might have a higher value of γ at higher densities than indicated by the calculations shown by means of curves 1 and 2 in Figure 11–1. This circumstance is suggested by the formation of pulsars during supernova explosions since the imploding stellar is likely to be $\simeq 1.4$ M$\odot$. At the present time two pulsars, the Crab pulsar and the Vela pulsar, are known to be associated with supernova remnants. The identifications indicate that neutron stars are formed during gravitational collapse. Curve 4 in Figure 11–1 shows an estimate for the nuclear equation of state that is based on an extrapolation of a semiempirical mass formula. If this equation of state were correct, relatively massive (that is, $M \sim 2$ M$\odot$) stable neutron stars could exist with central densities close to those inside normal nuclei. The interior densities of such neutron stars would be sufficiently low for superfluidity and superconductivity to be present throughout much of their mass.

Although the equation of state of very dense matter is not known, it is plausible to assume that

$$P \lesssim \rho c^2 \tag{11–27}$$

where ρc^2 is the total energy density. This condition is plausible because the speed of sound, which is

$$v_s = \left(\frac{dP}{d\rho}\right)^{1/2} \tag{11–28}$$

for relativistic as well as nonrelativistic matter should not exceed the speed of light c if the currently accepted views regarding causality are to remain valid. Therefore, no physically acceptable equation of state should cross the curve $P = \rho c^2$.

It is of interest to ask whether or not a stronger limit on physically plausible equations of state can be obtained. The pressure P and total energy density of a *noninteracting* gas of particles of number density n and mass m are

$$P = \tfrac{1}{3}nm \left\langle \frac{v^2}{\sqrt{1 - (v^2/c^2)}} \right\rangle \tag{11–29}$$

and

$$\rho c^2 = nm \left\langle \frac{c^2}{\sqrt{1 - (v^2/c^2)}} \right\rangle \tag{11-30}$$

respectively. The brackets indicate appropriate averages over the distribution of particles which is likely to be Maxwellian. Equations (11-29) and (11-30) show that for a noninteracting gas, the condition

$$P \leq \tfrac{1}{3}\rho c^2 \tag{11-31}$$

holds. The equality is approached in the limit of very relativistic velocities. The condition (11-31) is equivalent to the requirement that the trace of the energy momentum tensor [Equation (11-5)] be nonnegative. This condition, which would suggest that $P \lesssim \rho/\sqrt{3}$, is not correct if interactions between particles are included.

To illustrate how condition (11-29) can be violated for interacting particles, we consider the following simple model. The potential energy of interaction between two baryons is assumed to be of the form

$$\frac{g^2}{r} \exp\left(\frac{-r}{r_d}\right) \tag{11-32}$$

where g is a constant that is a measure of the strength of the interaction; r_d is the range of the interaction; and r is the distance between baryons. The interaction energy per unit volume is

$$V_{\text{int}} = \frac{n^2}{,\, 2} \int_0^\infty \frac{g^2}{r} e^{-(r/r_d)} 4\pi r^2 \, dr$$

$$= 2\pi g^2 r_d^2 n^2 \tag{11-33}$$

where $n^2/2$ is the number of interacting pairs of particles per unit volume. The corresponding kinetic energy density is proportional to $n^{4/3}$ if the particles are relativistic and therefore the interaction energy dominates the kinetic energy at very high densities. The pressure can be written as

$$P = -\frac{d \text{ (energy per particle)}}{d \text{ (volume per particle)}} = -\frac{d(\rho c^2/n)}{d(1/n)}$$

$$= n \frac{d\rho c^2}{dn} - \rho c^2 = (\gamma - 1)\rho c^2 \tag{11-34}$$

where γ is defined to be $(d \ln \rho / d \ln n)$. Equation (11-34) shows that the requirement of causality implies that $\gamma \leq 2$. Equation (11-33) shows that $\gamma \to 2$ at very high densities for the above-mentioned model of interacting particles, and therefore for this model the condition (11-31) is exceeded.

Cooling of Neutron Star

If the density of a neutron star is not sufficiently high for π mesons to be produced, cooling should take place by means of the Urca reactions.

$$n + n \rightarrow n + p + e^- + \bar{v}_e$$

$$n + p + e^- \rightarrow n + n + v_e.$$

The calculated rates of the Urca reactions can be used to deduce the rate of decrease of the core temperature. The rate of energy production necessary to maintain the Crab nebula is $\sim 10^{38}$ ergs/sec. Calculations of the cooling rates of neutron stars predict that a neutron star as old as the Crab nebula ($\sim 10^3$ yr) will have interior temperatures $\lesssim 10^8$°K and therefore not be an important *thermal* x-ray source. In the interior of a neutron star, the plasma density is so high that for temperatures $\lesssim 10^8$°K, photons cannot be excited thermally [see Equation (10–5)]. Calculated cooling rates are shown in Figure 11–2.

11–4   GRAVITATIONAL COLLAPSE AND BLACK HOLES

Black dwarfs and neutron stars represent stable end points of stellar evolution. Stars that have used up all their nuclear energy sources can remain indefinitely in these equilibrium states. However, stars that are sufficiently

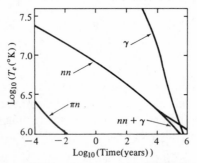

Figure 11–2   Cooling of neutron star. T_e is the effective temperature. The curve labeled γ represents the surface temperature as a function of time on the assumption that the cooling is caused solely by blackbody radiation from the surface. The curve *nn* gives the temperature on the assumption that cooling takes place exclusively by neutrinos emitted as a result of collisions between nucleons. The curve marked *nn* + γ describes the combined effect of curves *nn* and γ. The curve marked πn describes the much faster cooling that arise if π mesons are present in the interior of the neutron star.

Source: From Wheeler, J. 1966, *Ann. Rev. Astron. Ap.*, **4**, 393, by permission of Annual Reviews, Inc.

massive and do not lose mass cannot become white dwarfs or neutron stars. For such stars there exists no source of pressure that is sufficient to prevent gravitational collapse, and therefore they must undergo perpetual collapse unless it can be halted by rapid rotation.

Photons emitted from the surface of a collapsing star escape into space with their frequencies redshifted. As collapse continues, those photons that leave the surface of the star at some angle to the normal are bent into curved orbits. When the radius of the star is about 1.5 × the Schwarzschild radius, photons that are emitted tangent to the surface are trapped in a spherical cloud from which they gradually escape. This prediction means that an outside observer can see the rim of the star for an extended period of time. When the surface of a collapsing star reaches the Schwarzschild radius, only those photons emitted normal to the surface will escape. At higher densities, an outside observer will no longer receive photons from the collapsing star, and consequently he will no longer be able to communicate with an observer on the surface of the star. Although the existence of collapsed objects (called black holes) has not been demonstrated, it is likely that they are formed. They might be detected as the unseen companions of close binary systems. Since mass accretion into a black hole should produce x rays, it is likely that black holes are found among galactic x-ray sources.

CHAPTER 12
Galaxies

12–1 NORMAL GALAXIES

Galaxies are the largest and most massive systems of stars. Since the invention of the telescope, galaxies have been observed as faint, nebulous objects with measurable angular diameters. Although philosophers such as Immanuel Kant speculated that each of these nebulous objects were "island universes" similar to our own Milky Way, only in recent times has it been demonstrated that some of these observed nebulous objects were, indeed, large aggregates of stars. Historically, observations of novae inside these faint objects proved that they were very distant and consequently similar in size to our own Milky Way.

Galaxies are classified according to their appearances. The Hubble classification scheme for galaxies (Figure 12–1) defines three main classes: elliptical (E), spiral (S), and irregular (Irr). Recent studies have tended to attach much greater importance to galactic nuclei than Hubble's work did. Observations indicate that of the observationally preferred bright galaxies, approximately 75 percent are spiral galaxies, and approximately 20 percent are elliptical galaxies. The remaining galaxies are classified as irregular (Irr) galaxies.

The intrinsic luminosities and diameters of galaxies vary over a great range. Some dwarf galaxies are not much more luminous than globular clusters. On the other hand, giant elliptical galaxies or giant spiral galaxies such as the Andromeda Nebula (M31) have luminosities in excess of 10^{10} L$\odot$. The very large spread in intrinsic luminosities means that the average galaxy observed in a survey based on apparent magnitude is likely to be very different in type and luminosity from the average galaxy inside a given volume of space.

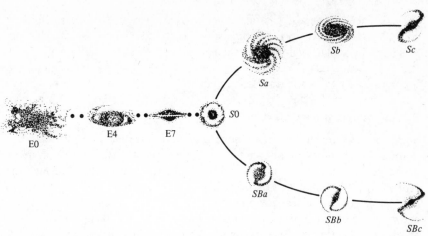

Sb

Sc

Sa

E0

E4

E7

S0

SBa

SBb

SBc

Figure 12–1   Hubble's classification scheme for galaxies.

Source: From EXPLORATION OF THE UNIVERSE, second edition, by George Abell. Copyright © 1964, 1969 by Holt, Rinehart and Winston, Inc. Reprinted by permission of Holt, Rinehart and Winston, Inc.

Elliptical Galaxies

The observed isophotes of elliptical galaxies indicate that they have rotational symmetry. Their diameters range from ~ 1–50 kpc, and their inferred geometrical forms are oblate spheroids (ellipsoids with two equal principal axes and a third smaller axis). Elliptical galaxies are subdivided according to the ratio of major to minor axis. The subdivisions are identified by the number $10(a - b)/a$, where a is the major axis and b the minor axis. It follows that E0 galaxies are nearly circular, while E7 galaxies are very flat. The surface brightnesses of elliptical galaxies are found to fall off from the axes according to a common empirical relation. This circumstance is of considerable importance in comparing the total fluxes from distant and relatively nearby galaxies.

The masses of elliptical galaxies range from 10^5–10^{12} M⊙. There is some observational evidence for the existence of a well-defined upper limit to their masses. Giant elliptical galaxies of similar luminosity are observed to be the most luminous galaxies in rich clusters of galaxies. This empirical result is of great importance in determining the extragalactic distance scale. Radio galaxies are frequently giant elliptical galaxies.

The observed colors and spectra of elliptical galaxies indicate that their stellar population is predominantly population II. Globular clusters are commonly observed, but OB type main-sequence stars are not present in appreciable numbers. Although very little dust or neutral gas is observed in elliptical galaxies, Hα emission and forbidden oxygen lines, which indicate the existence of ionized gas, are sometimes observed. Moreover, elliptical galaxies are observed to have some ultraviolet excesses. These excesses are predicted by stellar evolutionary calculations which show that a star should emit significant far ultraviolet radiation as a hot white dwarf before it becomes a normal white dwarf.

Spiral and Irregular Galaxies

Spiral galaxies are subdivided as normal or barred spirals. However, galaxies of intermediate types are known to exist. The arms of barred spirals originate at the ends of a luminous bar that is symmetrical about a plane through the center. The two main classes of galaxies are further divided according to relative size of the nuclear region, tightness of the winding of the spiral arms, and amount of patchiness of the spiral arms. Type *Sa* galaxies have a large amorphous nuclear region with tightly wound spiral arms. *Sc* galaxies have rather inconspicuous nuclear regions and very loose spiral arms. *Sb* galaxies are intermediate between *Sa* and *Sc* galaxies. *So* galaxies have a disk but no spiral arms.

Spiral galaxies unlike elliptical galaxies contain many young stars. Most of these young, massive OB type stars are associated with the spiral arms as are the HII regions that they produce. There is evidence for appreciable neutral gas and dust inside spiral galaxies.

The five main components of spiral galaxies are the nucleus, halo, disk, flat component, and spiral arms. Although most spirals have an observable nucleus, the luminosities of galactic nuclei vary widely. For certain spirals, notably Seyfert galaxies and N galaxies, the nucleus is very bright, and there is evidence of recent explosive events. The structure and content of the halo is similar to that of elliptical galaxies. Spiral galaxies also contain disk and flat stellar components as well as spiral arms that are embedded inside the flat component. The flat component contains the young stars and most of the neutral hydrogen.

Irregular galaxies lack rotational symmetry and are without recognizable patterns. Some irregular galaxies such as M82 contain large amounts of dust and show evidence for a recent explosive event. However, other irregular galaxies such as the Large Magellanic Clouds show little obscuration and can be easily resolved into stars.

Stellar Content

The observed color index of a galaxy is an indication of its classification. Elliptical galaxies are relatively red, irregular galaxies relatively blue, and spiral galaxies of intermediate color. As discussed above, this result is an indication of stellar content. The stellar population can be directly determined only if color-magnitude diagrams can be obtained. This is possible only for stars within galaxies of the local group. The most widely studied local group galaxies are the Large Magellanic Clouds, the Small Magellanic Clouds, and M31.

Several general conclusions can be made concerning the stellar content of galaxies. Elliptical galaxies contain old stars with little gas or dust and practically no young stars. The stellar population of spiral galaxies consists of several components.

Table 12–1

Component	Age (yr)	Metals
Halo	$\sim 10^{10}$	Weak-Normal
Disk	$\sim 10^{9}$	Normal
Flat-Spiral	$\sim 10^{8}$	Normal

The nucleus of M31 and our own galaxy consist primarily of old, metal-rich stars. All galaxies are observed to contain very old ($\sim 10^{10}$ yr) stars. This important result is consistent with the hypothesis that there was a preferred epoch for the formation of galaxies.

Diameter and Integrated Magnitude

In order to find the absolute magnitude of a galaxy, the brightness distribution and distance must be determined. The brightnesses of the extended outer regions of a galaxy is difficult to estimate because of their faintness. Systematic errors in finding the magnitudes of galaxies would invalidate cosmological conclusions and, therefore, are of considerable interest. Because the brightness of a galaxy decreases as a function of distance from the center, it eventually becomes less than the sky background. For this reason, it is necessary to define a diameter of a galaxy in such a manner that what we call the total luminosity is the total luminosity emitted from the specified diameter. There are two methods of defining the diameter of a galaxy. The first method assumes some standard intensity curve obtained from bright galaxies of similar type. The second method measures the size of a galaxy out to some well-defined surface brightness. If the stellar luminosity function and dust content are assumed known, the observed surface brightness can be related to the stellar density.

12–2   MASSES OF GALAXIES

The determination of the mass distribution and total masses of galaxies is of great importance in explaining their intrinsic properties. Moreover, most of the *known* mass is contained in galaxies. Since the overall behavior of space (that is, whether or not the universe is open or closed—see Chapters 1 and 13) is believed to be dependent on the average mass-energy density, knowledge of the total mass density of galaxies is of considerable cosmological importance. There are four basic methods for estimating the masses of galaxies. Although each of these methods of estimating mass is somewhat uncertain, it is hoped that their internal self-consistency is a measure of the overall uncertainty.

Rotation Curves

The most accurate method of determining the masses of spiral galaxies is by means of finding their rotation curves (that is, finding the rotational velocity as a function of distance from the nucleus). Narrow optical interstellar emission lines and 21-cm hydrogen absorption lines are used to determine rotational velocities. An example of a rotation curve is given in Figure 12–2.

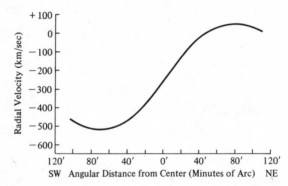

Figure 12–2   Rotation curve of M31 (Andromeda nebula). (Based on measurements of the radial velocities of emission nebulae in the spiral arms by N.U. Mayall.)

Source: From EXPLORATION OF THE UNIVERSE, second edition, by George Abell. Copyright © 1964, 1969 by Holt, Rinehart and Winston, Inc. Reprinted by permission of Holt, Rinehart and Winston, Inc.

In order to find the rotation curve, one must determine the linear scale factor which depends on the distance to the galaxy, the velocity of the center of the galaxy, and the angle between the rotation axis and the line of sight. The projection of the rotation axis along the line of sight is estimated by assuming that the rotation would be nearly circular if the galaxy were observed along the rotation axis. We assume that the gravitational force is balanced by centrifugal force. After the velocity curve has been measured, we look for a mass distribution that will fit the observed rotation curve.

Let us consider a simple model for a galaxy. Assume that a point mass M_1 is concentrated at the center, a mass M_2 is distributed with uniform density ρ over a sphere of radius R. The requirement that gravitational force balances centrifugal force implies

$$F_g = \frac{Gm_sM_1}{r^2} + \frac{4}{3}\pi Gm_s\rho r = \frac{m_sv^2}{r} \qquad (\text{if} \quad r < R)$$

$$F_g = \frac{Gm_s(M_1 + M_2)}{r^2} = \frac{m_sv^2}{r} \qquad (\text{if} \quad r > R)$$

(12–1)

where m_s is the mass of a star. If the mass of the nucleus M_1 is small, Equation (12–1) implies rigid body rotation out to some radius R. Because of Kepler's laws of motion, the velocity will decrease as $r^{-1/2}$ for larger distances from the center. More detailed theoretical models for rotation curves generally assume a spheroidal mass distribution.

The observed rotation curves of spiral galaxies show rigid body rotation out to some distance R from the nucleus. At distances greater than R, the velocity curves show approximately constant velocity until the outermost regions of the galaxy where the velocity must decrease as $r^{-1/2}$. The outermost regions of a galaxy are relatively faint and therefore difficult to observe. For this reason, the rotation curve method of determining the masses of galaxies can give only a lower limit to the mass. The masses of spiral galaxies found by means of rotation curves are 10^{10}–10^{11} M$\odot$ and have $M/L \sim$ 2–20. However, since only giant galaxies with many HII regions have observed rotation curves, the observed spiral galaxies are unlikely to be representative of all spiral galaxies.

Other Methods of Estimating Mass

A second method of estimating the masses of galaxies takes advantage of the existence of double galaxies. These galaxies have sufficiently small angular separations that it is plausible to assume that they are in bound orbits. It is further assumed that their orbits are circular and distributed with random orientation. Mean values for the masses of the major classes of galaxies can be found by measuring radial velocities and estimating distances.

If two galaxies of mass M_1 and M_2 (see Figure 12–3) are moving about each other in circular orbit of radius d, their relative velocity is

$$v = \sqrt{\frac{G(M_1 + M_2)}{d}}.$$

(12–2)

The y axis is chosen to be along the relative velocity v, and yz is the orbital plane. The component of v along the line of sight is

$$v_\parallel = v \sin \theta \sin \phi.$$

(12–3a)

From Equations (12–2) and (12–3a), we find

$$Rv_\parallel^2 = G(M_1 + M_2) \sin^3 \theta \sin^2 \phi.$$

(12–3b)

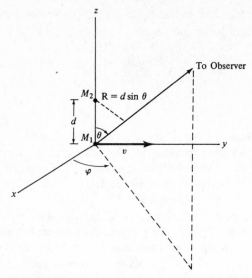

Figure 12–3 Two orbiting galaxies of mass M_1 and M_2 are separated by a distance d. v is the velocity of M_1.

The left-hand side of Equation (12–3b) can be determined since $v_\parallel$ can be measured and R can be estimated from the distance and angular separation of the galaxies. If a number of binary systems are observed with random orientation, we find

$$\overline{Rv_\parallel{}^2} = 0.3G(\overline{M_1 + M_2}). \tag{12–4}$$

The estimated mean mass of spiral galaxies that are in double systems is 4×10^{10} M$\odot$, and the mean M/L ratio is 3. The corresponding mean mass and M/L ratio of elliptical galaxies are 7×10^{11} M$\odot$ and 100, respectively.

A third method of estimating masses of galaxies is based on measuring the velocity dispersion in the central regions of the galaxies. If the masses and positions of the stars are denoted by M_i and r_i, respectively, and the M/L ratios are known, the measured velocity dispersion and brightness distribution can be used in conjunction with the virial theorem

$$\sum_i M_i v_i{}^2 = \frac{1}{2} \sum_i \sum_{j \neq i} \frac{GM_i M_j}{|\mathbf{r}_j - \mathbf{r}_i|} \tag{12–5}$$

to determine the mass. Although this method depends on a number of assumptions, it gives results that are consistent with the double galaxy method.

Many galaxies are found in rich clusters of galaxies. If the cluster is stable, then the virial theorem can be applied to the entire cluster, and mean values for the mass and M/L ratios can be found. This method of estimating mass gives mean values that are approximately ten times larger than indicated by other methods of determination. The corresponding M/L ratios are also greater by a factor of about ten. This rather puzzling result suggests either that clusters are not stable or that larger amounts of undetected mass are present.[1] The possible presence of large amounts of undetected mass is of great cosmological interest since the observed mass density is not sufficient to produce a closed universe (see Chapter 13).

12–3 SPIRAL ARMS

Much of the neutral interstellar gas, as well as many hot stars and HII regions, are confined to the spiral arms of a galaxy. The persistence of spiral structure against differential rotation is indicated by the fact that the spacing between spiral arms, which is used to classify spiral galaxies (*Sa, Sb, Sc,* and so on), is correlated with other characteristics of the galaxy such as its total mass and gas content. Since the inner regions of a galaxy rotate at a rate that is several times greater than the outer regions, the spiral patterns would change appreciably over a period of several rotations if spiral arms were material arms. This winding dilemma is removed if it is assumed that spiral arms represent wave patterns (density waves) rather than rotating material arms. The presence of density waves means that stars, which are in epicyclic motion, tend to aggregate into a spiral gravitational well that is, in turn, maintained by the excess of stars already in the gravitational well. Recent numerical experiments with models of the galaxy support the concept of density waves.

Observations of galaxies show that the spiral patterns are usually two-armed with the spiral arms trailing. However, multiple-armed structures often arise in the outer regions of spiral galaxies. Young, hot stars and HII regions occur in a narrow region in the arms. This circumstance suggests that star formation arises simultaneously in a restricted region. Dust lanes are generally found along the inner side of a bright spiral arm. The presence of density waves could lead to a compression of the gas near the minimum of the gravitational potential. Such a compression is likely to produce narrow dust lanes and may trigger star formation.

Although it appears likely that the phenomenon of spiral structure cannot be explained solely on the basis of rotating material arms, some galaxies, notably barred spirals, are likely to have material arms. Although the density wave theory of spiral structure is a very plausible theory, at the present time there is no well-understood physical mechanism for driving the wave patterns. Jeans instability in the outer regions of the galaxy is a possible mechanism.

---

[1] It is difficult to understand how clusters of galaxies could be unstable since the predicted time scale for their dispersal into space is less than the presumed lifetime of the constituent galaxies.

12-4   SEYFERT GALAXIES, N GALAXIES, AND THE INFRARED PHENOMENON

The infrared emission from a number of extragalactic sources has been found to be much greater than the total optical radiation from the most luminous galaxies. Most of the galaxies with very high infrared luminosities are Seyfert galaxies or N galaxies, which are distant radio galaxies with starlike nuclei. However, other exploding galaxies such as M82, which underwent an explosion in its nucleus $\sim 2 \times 10^6$ yr ago, are also known to be strong infrared sources. It appears that the phenomenon or phenomena that produce infrared galaxies are present at various power levels in the nuclei of most galaxies. This point of view is supported by the similarities between the infrared spectra of extragalactic nuclei including the nucleus of our own galaxy. The infrared spectra of a number of extragalactic sources are characterized by flux density maxima at $\sim 70\mu$ and sharp declines in flux density at longer wavelengths.

Seyfert galaxies are defined by (1) starlike nuclei, (2) forbidden emission line spectra that arise from ions of high excitation such as [OIII] and [NeIII] (the inferred electron densities are $\sim 10^4$ cm^{-3}, and the corresponding electron temperatures are $\sim 2 \times 10^{4\circ}$K), and (3) very broad ($\sim 1$–$5 \times 10^3$ km/sec) emission line spectra that include Balmer series lines.

Measurements of time variations in the flux density from Seyfert nuclei indicate that their diameters are very small (probably less than one light year). This conclusion is confirmed by recent angular size measurements of the nucleus of the closest Seyfert galaxy NGC 4151. The measurements establish an upper limit of 0.1 seconds of arc for its angular size, which implies that the linear dimension is less than 1 pc. The small linear dimensions of the nuclei of Seyfert galaxies make it improbable that thermal emission by dust grains is the principal mode of emission. Synchrotron radiation from relativistic electrons is a likely emission process; however, the sharp decline in the flux density at long wavelengths places important limitations on the classes of synchrotron model that are plausible.

Since approximately one percent of all spiral galaxies are Seyfert galaxies, it is likely that the Seyfert phenomenon lasts $\sim 10^8$ yr (that is, $\sim$ one percent of the Hubble time). The observed infrared luminosities of Seyfert galaxies are as high as 10^{45}–10^{46} ergs/sec. The luminosities and inferred ages of Seyfert galaxies imply that the total energy requirement is greater than 10^{60}–10^{61} ergs, which is the equivalent rest mass energy of $\sim 10^6$–10^7 M$\odot$. This represents an enormous expenditure of energy, and it has been suggested that physical processes totally different from those that have been investigated in the laboratory may take place inside galactic nuclei. The more conventional point of view is that this energy release is the result of gravitational collapse. It is possible but less likely that nuclear energy generation is the dominant energy source.

In Section 12–2 we showed how galactic rotation curves may be used to estimate the mass distributions inside galaxies. It is of considerable interest to estimate the masses of the nuclei of Seyfert galaxies. Rotation curves have been obtained for several Seyfert galaxies. Three separate components of ionized gas are discerned. These components include rotating gas in the vicinity of the nucleus, random velocity gas whose velocities often exceed the escape velocity from the nucleus, and a low-density component that shows the normal rotational velocity of the spiral arms. Upper limits of $\sim 4 \times 10^9$ M$\odot$ and $\sim 10^{10}$ M$\odot$, respectively, have been measured for two Seyfert nuclei, NGC 1068 and NGC 7469. Although these upper limits for the masses of the nuclei are sufficient for gravitational collapse to supply the required energy, an efficient mechanism of energy release is indicated.

12–5   RADIO GALAXIES

Certain galaxies are observed to be strong radio sources. It is widely believed that most radio emission from these radio galaxies is caused by the synchrotron radiation of relativistic electrons moving in magnetic fields. In extended radio sources, these magnetic fields are likely to be weak ($\sim 10^{-4}$– 10^{-5} G). However, much stronger magnetic fields may arise in compact radio sources or components of extended sources. At the present time, there is no unified theory for radio galaxies such as exists for stars. In particular, the possible relation between radio galaxies, active galactic nuclei, and quasistellar objects is not understood. Observations of radio galaxies indicate that they have complicated structures which include both extended and compact emission. Presumably the compact components are the energy sources for the extended objects, which are often wisps or jets and, therefore, likely to be the result of explosive events.

The strong and relatively distant ($z \equiv \Delta\lambda/\lambda \simeq 0.06$) radio source Cygnus A was the first radio galaxy to be identified. The most distant known radio galaxy is 3C295, which has a redshift of $z = 0.46$. Although it is undoubtedly true that many radio galaxies are far more distant than 3C295, only the distances to radio sources that can be identified with optical objects of known redshift can be determined. From a practical point of view, this means that the distances to galaxies with $z \geq 0.5$ cannot be measured. This circumstance in addition to our lack of knowledge with regard to their intrinsic nature places very severe limitations on the usefulness of radio galaxies for cosmology. Radio cosmology will be discussed further in Chapter 13.

Radio and Optical Properties

Cygnus A is a very intense radio source that consists of two extended radio components. The diameters of these components are about 50 kpc, and they are separated by about 150 kpc. A visible object that is approximately

1/10 of the diameter of the radio components is located between them. It consists of two plasma clouds that move in opposite directions or rotate about the nucleus with velocities of several thousand kilometers per second. Cygnus A was initially believed to represent two galaxies in collision. However, this interpretation does not explain the enormous energy output. A more likely interpretation of the phenomenon is that the nucleus of the galaxy exploded, sending gas jets at the same time in opposite directions. Recently, point sources (<1 seconds of arc) have been found at the centers of the radio emission. The presence of compact sources of radiation in Cygnus A suggests that an active energy source may still be present.

M82 is a relatively close irregular galaxy that recently underwent an explosion in its nucleus. Despite its irregular shape, it has two preferred orientations: one along the maximum longitudinal dimension of the galaxy and the other perpendicular to it. A system of filaments that emit primarily line radiation (strongest in Hα) is observed. These filaments move from the center at velocities of approximately 10^3 km/sec. The observed velocities and the dimensions of the filaments indicate that the explosion took place approximately 2×10^6 yr ago.

Virgo A(M87) is a giant elliptical galaxy, which is also an intense radio source. A jet is observed to extend from the central part of the galaxy. This jet, which is an intense emitter of ultraviolet and x-ray radiation, consists of several highly polarized condensations. The existence of a jet is indicative of an explosive event.

Because many radio galaxies are very distant and their optical counterparts faint, accurate radio positions are necessary if reliable identification is to be possible. Typically radio positions to an accuracy of about 15 seconds of arc in each coordinate are necessary for identification. At the present time, the radio positions of nearly 10^3 radio sources are known to the required accuracy. About one half of these sources can be identified with galaxies or quasistellar objects.

Normal galaxies are usually weak radio sources ($L_R \lesssim 10^{38}$ ergs/sec). On the other hand, radio galaxies, which are by definition strong radio emitters, have radio luminosities that range from 10^{40}–10^{44} ergs/sec. Most elliptical galaxies are weak radio sources. Consequently, only a few have been detected as radio emitters. Spiral galaxies such as our own galaxy are generally somewhat stronger radio sources ($L_R \sim 10^{38}$ ergs/sec). Seyfert galaxies, which are spiral galaxies with active nuclei, are strong radio sources. The strongest radio galaxies tend to be giant elliptical galaxies, D galaxies which are single objects in a fainter extended envelope, N galaxies which have bright stellar nuclei in a faint envelope and therefore resemble quasistellar objects, dumbbell galaxies, which have double nuclei inside a single envelope, or certain irregular galaxies. Approximately 30 percent of radio galaxies are found in rich clusters of galaxies.

Polarization

The radiation from radio galaxies is generally found to be linearly polarized. This result is consistent with the point of view that the radio emission is primarily synchrotron radiation. The measured degree of linear polarization ranges from ~0.5 percent to ~20 percent and is generally less at longer wavelengths. The rather low degree of linear polarization observed in most sources is likely to be the result of randomly oriented magnetic fields and depolarization due to Faraday rotation inside the source.

Angular Sizes

Radio galaxies usually have a fair amount of structure, and their angular sizes are generally <1 minute of arc. The most common structure for radio galaxies has most of the radio emission coming from the sources of comparable but not equal luminosity. About 30–50 percent of radio galaxies are contained in this general classification. About 10 percent are core-halo objects, while the remaining radio galaxies have more complicated structures. When radio sources are observed at high angular resolution, knots of high brightness temperature are observed. Nearly 30 percent of radio sources have some structure that is unresolved at 1 second of arc. Moreover, some radio galaxies have structure that is unresolved even at an angular resolution of 0.02 seconds of arc. Angular size and flux measurement provide some evidence that the brightness temperatures of radio galaxies are $\lesssim 10^{12}°K$.

Radio Spectra

The radio spectra of most sources can be represented fairly well by a simple power law

$$F_\nu \propto \nu^{-\alpha} \tag{12-6}$$

where α, the spectral index, is typically ~0.8. The spectrum and high effective brightness temperatures of radio sources indicate that the radio emission is nonthermal. The existence of simple power law spectra is most generally true for wavelengths in the range 200 cm $\geq \lambda \geq$ 20 cm. The radio spectra of some sources show low-frequency cutoffs. At shorter wavelengths, the spectra of some sources deviate from a simple power law. The observed curvature is usually negative [that is, F_ν decreases more rapidly than predicted by Equation (12-6)]. However, the flux from some radio sources increases at short wavelengths.

Compact radio sources tend to have flatter spectra than other sources and to show short-term variability at centimeter wavelengths. Such variability indicates continued injection of high-energy particles. Low-frequency cutoffs are also characteristic of compact sources. The rapid decline in the vicinity of these low-frequency cutoffs is too rapid to be explained on the basis of a

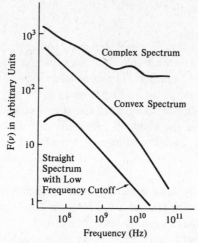

Figure 12-4 Typical spectra of radio galaxies are shown in arbitrary units.

peculiar distribution of electrons. Although free-free absorption and the refractive index of the plasma may be important, synchrotron self-absorption is the most likely cause of the low-frequency cutoffs. Figure 12-4 shows several basic types of radio spectra.

Energy Requirements

Estimates of the total energy necessary to form a radio galaxy are of considerable interest. The required energy is obviously enormous, and it has even been suggested that known physical processes are inadequate to explain the total energy release. Much of the energy associated with radio galaxies is in the form of relativistic particles and magnetic fields.

We assume that the radio luminosity is due to synchrotron radiation from relativistic electrons that are distributed isotropically and uniformly over the volume of the radio galaxy. The number of relativistic electrons of energy E per unit energy interval is assumed to be

$$N(E) = KE^{-\gamma} \quad E_1 < E < E_2$$
$$N(E) = 0 \qquad E < E_1 \quad \text{or} \quad E > E_2 \tag{12-7}$$

where K and γ are constants. The above form of the distribution function $N(E)$ is taken because it leads to a natural explanation of the radio spectra of most radio galaxies and also because the corresponding distribution function for cosmic-ray particles has this form. Equation (12-7) implies that the total energy in relativistic electrons is

$$E_e = \int_{E_1}^{E_2} EN(E) \, dE = \frac{K[E_2^{2-\gamma} - E_1^{2-\gamma}]}{(2 - \gamma)} \quad (\gamma \neq 2). \tag{12-8}$$

If synchrotron emission is the dominant energy loss mechanism for electrons, the rate of energy loss per electron is

$$\frac{dE}{dt} = -AB_{\perp}^{2}E_{\text{GeV}}^{2} \qquad \text{ergs/sec} \tag{12-9}$$

where $A = 6 \times 10^{-9}$, and $B_{\perp}$ is the component of the magnetic field perpendicular to the velocity of the electron. It follows from Equations (12–7) and (12–9) that the radio luminosity is

$$L_R = -\int_{E_1}^{E_2} N(E)\frac{dE}{dt}\,dE$$

$$= \frac{AKB_{\perp}^{2}}{(3 - \gamma)}\left[E_2^{(3-\gamma)} - E_1^{(3-\gamma)}\right] \qquad (\gamma \neq 3). \tag{12-10}$$

The radio luminosity is approximately 10^{44} ergs/sec for a luminous radio galaxy such as Cygnus A.

Equations (12–8) and (12–10) imply

$$E_e = \frac{(3 - \gamma)}{(2 - \gamma)}\frac{(E_2^{2-\gamma} + E_1^{2-\gamma}L_R)}{(E_2^{3-\gamma} - E_1^{3-\gamma})B_{\perp}^{2}A} \qquad (\gamma \neq 2 \text{ or } 3). \tag{12-11}$$

A plausible theory for the origin of radio galaxies asserts that they are formed as a result of streams of relativistic particles (mostly electrons and protons) that are ejected from the nuclear regions of the galaxy and subsequently interact with existing intergalactic matter and magnetic fields. The very large linear dimensions of giant radio galaxies ($\sim 10^5$ pc) indicates that their lifetimes exceed 10^6 yr. Their present radio luminosities and likely minimum ages indicate an energy release of at least 10^{58} ergs in the form of relativistic electrons. The energy released in the form of relativistic protons may be much higher.

If the magnetic field is assumed to be constant throughout a radio source, the magnetic energy is

$$E_m = \frac{B^2}{8\pi}V \tag{12-12}$$

where V is the volume of the source. An estimate for the total magnetic energy can be formed from Equations (12–11) and (12–12) if it is assumed that the total energy $E = E_e + E_m$ is minimized with respect to changes in the magnetic field. This estimate leads to magnetic fields of $B \sim 10^{-4}$ G and total magnetic energies that are comparable to the energy in the form of relativistic electrons. In making this estimate, one assumes that each electron radiates at its critical frequency defined in Equation (9–19) and then expresses E in terms of the cutoff frequencies corresponding to E_2 and E_1, respectively, and the magnetic field B.

Energy Losses and the Continuity Equation

The simplest model for a radio galaxy assumes that it is the result of recurring bursts of high-energy particles that are confined to magnetic clouds (plasmoids). It is assumed that the magnetic flux remains constant during expansion and that the expansion is nearly adiabatic. These assumptions make it possible to construct simple models. On the average, extended radio galaxies are expected to be older than compact sources.

We wish to show that during adiabatic expansion, the energy of a relativistic particle will vary as $1/r$, where r is the radius of the gas cloud. The first law of thermodynamics implies

$$dQ = dE + P\,dV$$

$$= dE + \frac{1}{3}\left(\frac{E}{V}\right) dV = 0 \qquad (12\text{--}13)$$

where we have used the relation $P = \frac{1}{3}(E/V)$, which holds for a noninteracting relativistic gas. Since $V \sim r^3$ the relation $E \sim 1/r$ follows from Equation (12–13).

Table 12–2 lists the most important energy losses that relativistic electrons are expected to encounter inside radio galaxies as well as how these energy losses are expected to vary with the linear dimension of the source. The assumption of magnetic flux constancy implies $B \sim 1/r^2$. The dimensional relations shown in the fourth column of Table 12–2 were derived under the assumption that the energy losses are primarily due to expansion. In Table 12–2, v is the expansion velocity of the cloud (assumed constant), and E_r is the radiation energy density. It should be emphasized that statistical acceleration of particles, as well as energy loss, may take place in a radio source. The dimensional factors shown in Table 12–2 indicate that inverse Compton scattering and synchrotron radiation will dominate for very compact sources, but that expansive losses will be most important for extended sources.

Table 12–2   ENERGY LOSSES

Type	$-dE/dt$	Dimensional Factors	
Expansion	$(v/r)E$	$r^{-1}E$	r^{-2}
Synchrotron radiation	$10^{-3}B_\perp{}^2(E/mc^2)^2$	$r^{-4}E^2$	r^{-6}
Inverse Compton*	$2 \times 10^{-14}E_r(E/mc^2)^2$	$r^{-8}E^2$	r^{-10}
Bremsstrahlung	$8 \times 10^{-16}n_eE$	$r^{-3}E$	r^{-4}
Nuclear collisions	$2.5 \times 10^{-16}n_eE$	$r^{-3}E$	r^{-4}
Ionization		r^{-3}	r^{-3}

* Here we assume that E_r is due to synchrotron radiation.

If the energy losses are such that many collisions are required to reduce the energy of the particle by a significant fraction, a continuity equation in energy space can be found for $N(E, t)$, which is defined to be the number of electrons per unit energy between E and $E + dE$ at time t. This condition is satisfied for synchrotron radiation, since the energy loss involves the emission of many low-energy photons unless the magnetic field is very large. The equation for continuity in energy space is

$$\frac{\partial N}{\partial t} (E, t) + \frac{\partial}{\partial E} J(E, t) = Q(E, t) \tag{12-14}$$

where $J = N(E, t)(dE/dt)$ is the flux of electrons entering the energy interval dE between E and $E + dE$, and $Q(E, t)$ represents the possible sinks and sources for the relativistic electrons. Before Equation (12–14) can be solved, dE/dt and $Q(E, t)$ must be specified. We emphasize that Equation (12–14) is inadequate if an electron gives up a large fraction of its energy to one photon such as is the case in high-energy Compton scattering.

If synchrotron losses dominate and there are no sources or sinks of particles, Equation (12–14) becomes

$$\frac{\partial N}{\partial t} (E, t) - AB_\perp^2 \frac{\partial}{\partial E} [E^2 N(E, t)] = 0. \tag{12-15}$$

Before we can solve Equation (12–15), we must determine how E varies with t. Solving Equation (12–9), we find that for an electron of initial energy E_0, the time dependence of the energy is

$$E = \frac{E_0}{1 + AB_\perp^2 E_0 t}. \tag{12-16}$$

Equation (12–16) shows that even an electron of infinitely high initial energy E_0 would have its energy reduced to $E = 1/(AB_\perp^2 t)$ after a finite time t.

Models for radio sources can be obtained by solving Equation (12–14) subject to various assumptions regarding electron injection energy losses and particle acceleration. A particularly interesting simple model assumes that beginning at $t = 0$, a constant spectrum $qE^{-\gamma}$ of electrons with an isotropic velocity distribution is injected into a region initially devoid of relativistic electrons. It can be shown that this model leads to the solutions

$$N(E, t) = 4\pi q E^{-\gamma} \qquad \left(E \ll \frac{1}{AB_\perp^2 t} \right)$$

$$= \frac{2\pi q E^{-(\gamma+1)}}{AB_\perp^2 (\gamma - 1) t} \qquad \left(E \gg \frac{1}{AB_\perp^2 t} \quad \text{and} \quad \gamma \neq 1 \right). \tag{12-17}$$

Equation (12–17) predicts a break in the electron energy spectrum that leads to a break in the observed photon spectrum. This simple model for a radio galaxy provides a possible explanation for the curvature observed in the spectra of certain radio galaxies such as Cygnus A. If this model is correct, then the age of the source is related to the frequency at which the break in the spectrum arises.

Synchrotron Self-Absorption (Heuristic Discussion)

The flux density F_ν of a radio source generally increases with decreasing frequency. However, at sufficiently low frequencies, the synchrotron spectrum will be markedly changed by synchrotron self-absorption, and F_ν will decrease with decreasing frequency. If the intensity of synchrotron radiation is very high, the cutoff due to synchrotron self-absorption will occur at frequencies sufficiently high to be observed.

The following heuristic argument can be used to explain the general behavior of synchrotron self-absorption. The cutoff frequency for synchrotron self-absorption is assumed to arise when the surface brightness temperature T_b of the source is related to the flux density F_ν by means of the expression

$$F_\nu = 2kT_b \frac{\nu^2}{c^2} \Delta\Omega \qquad (12\text{–}18)$$

where $\Delta\Omega$ is the solid angle subtended by the source. We assume that electrons of energy E radiate at their critical frequency. Therefore, we have

$$kT_b \sim E \propto B^{-1/2}\nu^{1/2}. \qquad (12\text{–}19)$$

Equations (12–18) and (12–19) lead to the expression

$$F_\nu \propto B^{-1/2}\nu^{5/2}. \qquad (12\text{–}20)$$

More exact calculations based on the theory of synchrotron radiation verify the correctness of relation (12–20).

Inverse Compton Effect

For the case of Compton scattering of a photon by an electron that is at rest, the energy of the scattered photon is less than that of the incident photon. The opposite is true for inverse Compton scattering, which involves the scattering of a photon with a relativistic electron. In such collisions, the mean energy transferred by an electron of energy γmc^2 to a photon is approximately

$$\overline{\Delta E} = \gamma\bar{\varepsilon}' = \gamma^2\bar{\varepsilon} \qquad (12\text{–}21)$$

where $\bar{\varepsilon}'$ and $\bar{\varepsilon}$ are the mean energies of the photon in the frames of reference moving with the electron and with the observer, respectively. Equation (12–21) makes it clear that collisions of high-energy electrons with relatively low-energy photons (for example, 3°K background radiation) can lead to the production of x rays.

The rate of energy loss of an electron caused by the inverse Compton effect is

$$-\frac{dE}{dt} = + n_{ph}\sigma_T c\,\overline{\Delta E} \simeq + \sigma_T c \gamma^2 E_r \qquad (12\text{–}22)$$

where σ_T is the Thomson cross section; n_{ph} the photon number density; and $E_r = n_{ph}\bar{\varepsilon}$ is the mean radiation energy density. Equation (12–22) remains valid so long as $\bar{\varepsilon}\gamma < mc^2$. The rate of energy losses caused by inverse Compton scattering are proportioned to γ^2 of the electron just as in the case of synchrotron radiation (magnetobremsstrahlung). It is important to recognize that inverse Compton scattering is equivalent to the bremsstrahlung of electrons in the electromagnetic field of thermal radiation.

12–6 QUASISTELLAR OBJECTS (QSO)

Basic Properties

The early determinations of the positions of radio sources were not sufficiently accurate to make reliable identifications of radio sources with their probable optical counterparts. However, the subsequent determination of the positions of radio sources to several seconds of arc made reliable identifications feasible. A number of strong radio sources were identified with faint, starlike images on photographic plates. The radio sources 3C48 and 3C273 whose apparent visual magnitudes are $m_v = 16$ and $m_v = 12.6$, respectively, were the first two QSOs discovered. Broad emission lines were observed from these objects at wavelengths that did not correspond to those previously observed in normal objects such as stars. These broad emission lines were subsequently identified with redshifted emission lines such as are commonly observed from planetary nebulae. The measured redshifts of 3C48 and 3C273 were found to be $z = \Delta\lambda/\lambda = 0.367$ and 0.158, respectively. At the present time, nearly 200 QSOs have been identified, and it is believed that more than 10^3 QSOs exist in the radio source catalogs. It is now known that a class of QSO that does not emit a detectable amount of radio radiation is $\sim 10^2$ times more frequent that those associated with strong radio sources. This result indicates that at least 10^5 QSOs are likely to exist.

The basic observable properties that define QSOs as a distinct class of astronomical objects are

1. starlike objects often identified with radio sources known as quasars (QSS);
2. large redshifts of spectral lines;
3. broad emission lines (absorption lines sometimes present);
4. large ultraviolet and infrared excesses;
5. variable light.

On the basis of known optical properties, it has been suggested that QSOs are related to Seyfert galaxies and N galaxies. The radio properties of QSOs suggest that they may be related to radio galaxies. However, at the present time evidence regarding the intrinsic nature of QSOs is lacking.

Continuum Radiation

The emission line radiation observed from QSOs indicates that the emission line spectrum arises from regions in which the electron temperature T is $\sim 2 \times 10^{4}$°K. However, the general shape of the continuum spectrum indicates that it does not consist of thermal radiation from a plasma.

QSOs emit much more energy in the ultraviolet than normal stars and have large infrared excesses. Figure 12–5 shows the positions of quasistellar objects in the color-color diagram. Large short-term (weeks–years) time variations are observed in the continuum radiation of many QSOs but not from the emission line radiation. These observations indicate that the emission line radiation and continuum radiation arise from separate regions of space.

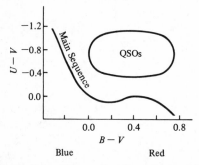

Figure 12–5   The regions of the color–color diagram occupied by QSO and main sequence stars. U, B, and V are the magnitudes as measured with the standard ultraviolet, blue and visual filters; respectively.

Both QSOs and radio galaxies show wide ranges of intrinsic flux levels. If the observed redshifts of QSOs are of cosmological origin, then the luminosities of quasars are 10^{44}–10^{46} ergs/sec which indicate that they are often 100 times more luminous than normal galaxies. Their infrared luminosities may be even higher. It is of great interest to estimate the total energy necessary to explain the QSO phenomenon. If it is assumed that the lifetime τ of a QSO is $\tau \gtrsim R/c$, where c is the speed of light and R the largest linear dimension, we can place a lower limit on the total energy requirements by determining the distances and angular diameters.

Radio interferometer measurements show that most QSOs are double (or multiple) objects. The separation between components often exceed 100 kpc if the redshifts of QSOs are assumed to be of cosmological origin. Moreover, many QSOs have very extended single components. For example, the QSO 3C273 is observed to have a bright starlike optical source and a faint extended optical jet. On the other hand, at radio wavelengths the extended jet is relatively bright as compared to the point object. The large dimensions associated with some components of QSOs indicate that more than 10^{60} ergs is necessary to explain the QSO phenomenon if they are at cosmological distances. The morphology of QSOs suggests that their compact components serve as the energy sources for the more extended components.

Synchrotron radiation is the most likely physical process responsible for the continuum radiation from QSOs. However, inverse Compton radiation and coherent plasma oscillations may also be important. The synchrotron process is favored because it can explain the spectrum and linear polarization of the observed radiation. The flux density F_ν from a QSO is typically proportional to $\nu^{-\alpha}$ with $\alpha \sim +0.7$ at radio wavelengths such that $\lambda \gtrsim 10$ cm. However, at centimeter wavelengths the spectra of QSOs (and some very bright radio galaxies) show significant curvature. It has been suggested that the observed curvature in the spectra of QSOs may be a consequence of synchrotron self-absorption.

The existence of radio bursts in QSOs at centimeter and millimeter wavelengths provides evidence that the dimensions of some components of QSOs are very small. The theoretical model that assumes a radio burst is due to a rapidly expanding cloud (or clouds) of high-energy particles is able to explain the basic behavior of the time variations of QSOs.

It is sometimes assumed that because time variations of the continuum radiation takes place in months or even days, the physical size of the emitting region must be less than c times the time scale for variations. However, this conclusion is not generally correct, since the emitting surface may appear to be moving faster than the speed of light (see discussion below).

It follows from Equations (9–15) and (12–22) that inverse Compton scattering will dominate synchrotron radiation for an isotropic distribution of electrons if

$$\frac{E_r}{(B^2/4\pi)} < 1 \qquad (12\text{–}23)$$

where E_r is the energy density of radiation. For an anisotropic distribution of particles, condition (12–23) becomes

$$\frac{E_r}{[(B^2 \sin^2 \theta)/4\pi]} < 1 \qquad (12\text{–}24)$$

where θ is the angle between the particles and the magnetic field. Conditions (12–23) and (12–24) show that if QSOs are at cosmological distances, their magnetic fields must be relatively high, since the density of radiation is very high. The presence of high magnetic fields implies short radiative lifetimes for the relativistic particles and therefore suggests that particles must be continuously accelerated inside QSOs. Similar arguments are likely to hold for Seyfert nuclei.

Emission and Absorption Line Spectra of QSOs

Quasistellar objects are characterized by very broad (20–30 Å) emission lines (both permitted and forbidden transitions are observed). The observed spectral lines come from ions with a wide range of ionization potential. The electron temperatures, densities, and chemical compositions inferred for QSOs are similar to those inferred for planetary nebulae. However, the equivalent widths of H recombination lines are less than predicted for the inferred electron temperature. It is likely that the radiation field is mostly nonthermal.

In diffuse nebulae and presumably also in QSOs, H and He lines are produced by the recombination of ions that were photoionized by ultraviolet radiation from a hot central source. Since the continuum radiation is observed to be time variable in a number of QSOs, the emission line strengths should also change. The amount of change should give us a means of testing models of QSOs, since the time dependence of the emission lines will depend on the properties of the nebula responsible for the emission lines as well as the time dependence of the ionizing radiation.

Electron collisions with atoms or ions are responsible for the observed forbidden lines. Low-lying metastable states are excited by electron collisions. If the density is sufficiently low, the decay of these metastable states will take place by means of radiation rather than collisional deexcitation. The occurrence and relative strengths of forbidden transitions allow one to estimate the densities and temperatures inside the emitting region. Typical number densities and temperatures are $n_e \sim 10^5$ and $T \sim 2 \times 10^{4\circ}$K, respectively. Although some small variations in the emission line strengths of QSOs have been observed, these variations in line strength are very much less pronounced than those observed from the continuum radiation.

Absorption lines are observed in the spectra of some QSOs. Moreover, multiple redshift systems are found in QSOs with rich absorption line spectra. Most absorption line redshifts are less than the corresponding redshifts of emission lines. This result suggests that the absorbing matter is either closer than the region containing the emission lines or moving toward us. However, some absorption redshifts occur with slightly higher z than the corresponding emission line redshifts. The origin of absorption lines in QSOs remains a mystery. It has been suggested that they may arise in gas clouds ejected from QSOs, intergalactic matter, or galaxies that lie along the line of sight.

Nature of Quasistellar Objects

The nature of quasistellar objects remains an enigma. At the present time, it is not known if QSOs are related to other astronomical objects such as galaxies or clusters of galaxies. The possible association of QSOs with clusters of galaxies is of particular interest, since approximately 30 percent of radio galaxies are in rich clusters.

The origin of the redshifts of QSOs is still uncertain. Three hypotheses regarding their origin have been advanced. The first and most widely held hypothesis asserts that the redshifts are of cosmological origin. If this is the case, then QSOs must be at very large distances, and consequently their small angular diameters imply that huge amounts of energy must be released inside very small regions of space. Before the cosmological hypothesis can be accepted, it is necessary that QSOs be identified with galaxies or intergalactic matter, which is known to be at cosmological distances. There does not appear to be a simple magnitude-redshift relation for QSOs. This result is not surprising, since the radio and even optical flux levels have a large scatter.

A second hypothesis is that the redshifts are of Doppler origin. The absence of blueshifts indicates that this hypothesis is plausible only if the QSOs were originally ejected from a relatively nearby region such as the galactic center. On the other hand, upper limits on the proper motions of QSOs indicate that their distances are greater than 200 kpc. A very large ($\gtrsim 10^{62}$ ergs) release of energy would be required to produce the large number of QSOs ($\sim 10^5$) and maintain their flux levels for the time necessary for them to reach their present positions in space. Another difficulty with the Doppler origin of QSOs arises because radio source counts are not easily understood if QSOs are the result of a local explosion.

A third hypothesis concerning the origin of QSO redshifts asserts that they are gravitational. Suppose that a neutron star or collapsed object of about 1 M$\odot$ is surrounded by a plasma shell such as might be assumed responsible for the emission line spectra of QSOs. If W is the measured linewidth and $z \equiv \Delta\lambda/\lambda$ is the redshift, the width of the shell is

$$\Delta R = \frac{W}{\lambda z} R \qquad (12\text{--}25)$$

where R is the radius of the emitting region. The measured value of $W/\lambda z$ is ~ 0.1 for the QSO 3C273. Since R is $\simeq 10^6$ cm for a 1-M$\odot$ neutron star, ΔR must be $\sim 10^5$ cm for 3C273. The electron temperature in the line emitting region is $\sim 10^{4\circ}$K, and the measured flux is $F(H\beta) \sim 3 \times 10^{-12}$ ergs-sec^{-1}-cm^{-2}. Moreover, the Hβ emission is

$$\varepsilon(H\beta) \simeq 10^{-25} n_e^2 \left(\frac{10^4}{T}\right) \quad \text{ergs-sec}^{-1}\text{-cm}^{-3}. \qquad (12\text{--}26)$$

Equations (12–25) and (12–26) lead to the equation

$$10^{-25}R^2 \, \Delta R n_e^{\,2} = 3 \times 10^{-12}D^2 \qquad (12\text{–}27)$$

where D is the distance. Since proper motion measurements show that the distance to 3C273 is >10 pc, it follows immediately from Equation (12–27) that the densities required to explain the existence of forbidden lines, such as those of [OIII] whose presence implies $n_e < 10^6$ cm^{-3}, are much too low for the shell to emit the required flux level.

More generally, it can be shown that redshifts as large as $z \geq 2$, such as are observed for QSOs, cannot arise from the surface of a nonrotating collapsed object. As discussed in Section 11–4, very few photons can be observed from a nonrotating object whose radius is $1.5 \times$ the Schwarzschild radius. This result implies that redshifts >0.6 should not be observable from such objects. Therefore, it is unlikely that QSO redshifts are caused by gravitation unless the collapsed object is rotating rapidly.

Time variations in the flux density have been observed from a number of QSOs at centimeter and millimeter wavelengths. The angular dimensions of these variable sources, which are compact components of larger sources, are always very small (<0.1 minute of arc). Moreover, their spectra show positive curvature (that is, their flux densities increase as a function of frequency). For this reason, it is likely that synchrotron self-absorption is important in these sources. The general properties of the observed time variations, which are illustrated in Figure 12–6, can be understood if it is

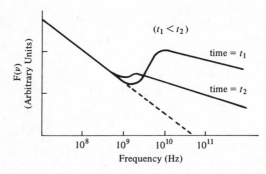

Figure 12–6 Typical time dependence of the flux density from variable radio source.

assumed that the radiation is caused by synchrotron radiation emitted by an expanding cloud of relativistic particles.

At wavelengths such that the synchrotron emission is optically thick, the flux density is

$$F(\nu, t) \propto B^{-1/2}(t)\theta^2(t)\nu^{5/2} \qquad (12\text{–}28)$$

where $\theta(t)$ is the angular size of the expanding component at time t. Equation (12–28) and the condition of magnetic flux constancy imply

$$F(v, t') = F(v, t)\left[\frac{r(t')}{r(t)}\right]^3 \qquad (t' > t) \qquad (12\text{–}29)$$

if the expansion is uniform, and the source remains optically thick at time t'.

It follows from Equation (9–21) that the corresponding time variation of the flux density at wavelengths for which the radiation is optically thin is

$$F(v, t') = F(v, t)\left(\frac{r(t')}{r(t)}\right)^{-2\gamma} \qquad (12\text{–}30)$$

where $r(t)$ is the radius of the source at time t, and $\alpha = (\gamma - 1)/2$ is the spectral index of the radiation. Equations (12–29) and (12–30) show that the flux density increases as a function of time at wavelengths for which the radiation is optically thick and decreases at wavelengths where the source is optically thin. The predicted behavior of the radiation at intermediate wavelengths is shown in Figure 12–6. The frequency at which the flux density is maximum moves to longer wavelengths as a function of time. Moreover, if the magnetic field has a preferred orientation, an increase in the degree of linear polarization and a rotation of plane of polarization are predicted as the source becomes optically thin.

The time variations from QSOs typically occur over periods of months. These rapid time variations suggest that the linear dimensions of the varying component is very small. It might seem plausible to conclude that the maximum extent of the varying component is $R = c\tau$, where τ is the characteristic time scale for the variability, and c is the speed of light. However, as will be shown below, the relativistic expansion of an optically thick source can produce an apparent rate of increase of angular size that is increased by a factor

$$\gamma = \left(1 - \frac{v^2}{c^2}\right)^{-1/2}$$

over the corresponding light travel time.

The following simple model illustrates how a cloud expanding at a relativistic velocity can appear to expand at a velocity greater than c. Assume that a spherical shell, which is produced by an explosion at point O' (see Figure 12–7) is expanding at a uniform velocity v. The problem is to determine the apparent perpendicular velocity of the visible envelope. We let $t = 0$ when a light signal emitted from O' and coincident with the explosion arrives at the observer O. This implies that the time of the explosion is $t_e = -l/c$, where l is the distance between the point of the explosion and the observer The points on the visible shell can be labeled by the distance R

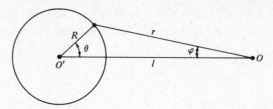

Figure 12–7 Sphere centered at O' appears to expand faster than c as viewed by observed at position O.

from O' and the angle θ between the time of observation t_0 at O and the corresponding time of emission t' at the point (R, θ) is

$$t_0 - t' = \frac{r}{c} \sim \frac{l}{c} - \frac{R \cos \theta}{c} \tag{12–31}$$

where r is the distance between point (R, θ) and O. It has been assumed that $R \ll l$. Since v is the velocity of the shell, it follows that

$$t' - t_e = \frac{R}{v}. \tag{12–32}$$

Eliminating t' from Equations (12–31) and (12–32) and using the expression $t_e = -l/c$, we find

$$R = \frac{vt_0}{1 - (v/c) \cos \theta}. \tag{12–33}$$

It follows from Equation (12–33) that the velocity perpendicular to the expanding shell is

$$u_\perp = \frac{dR}{dt} \sin \theta = \frac{v \sin \theta}{1 - (v/c) \cos \theta}. \tag{12–34}$$

The corresponding apparent rate of change of the angular diameter is

$$\frac{d\varphi}{dt} = \frac{v \sin \theta}{R(1 - (v/c) \cos \theta)}. \tag{12–35}$$

It can be readily shown that the apparent perpendicular velocity and rate of angular diameter are maximum when $\cos \theta = v/c$. Consequently, the maximum apparent perpendicular velocity is

$$u_{\perp(\text{max})} = \frac{v}{\sqrt{1 - (v^2/c^2)}} \tag{12–36}$$

which may exceed c.

At the present time, the problem of the true nature of QSOs remains unsolved. It is clear that the origin of their redshifts must be understood before realistic physical models for QSOs can be developed. Several QSOs have been discovered that are in the same directions as clusters of galaxies and in addition have the same redshifts as the cluster members. This circumstance would appear to support the point of view that QSO redshifts are of cosmological origin. However, it is not clear that this result should be generalized to include all QSOs. On the other hand, measurements of the change in the angular size of another QSO indicates an apparent change in the angular diameter that would correspond to a velocity 10 times the speed of light if the QSO is at a cosmological distance. Although these measurements may be interpreted on the basis of arguments such as we have described above, they may also be explained if QSOs are much closer than indicated by their redshifts.

Recent observations show that some objects that have been classified as QSOs are associated with galaxies. This circumstance suggests that the QSO phenomenon is associated with violent events in galactic nuclei. N galaxies and Seyfert galaxies appear to be classes of objects that represent a transition between normal galaxies and QSOs.

CHAPTER 13
Cosmology

13–1   MEASUREMENT OF HUBBLE CONSTANT AND DECELERATION PARAMETER

The classical problem of cosmology is that of discriminating among cosmological world models (called Friedmann models) on the basis of the observed kinematics of galaxies. It was pointed out in Chapter 1 that an accurate determination of the Hubble constant H, which measures the present rate of expansion of the universe, and q, the deceleration parameter, would make such a discrimination possible. Unfortunately, accurate values of H and q have not yet been determined.

In addition to their systematic recession, galaxies have random components of velocity. These random velocities, which are typically 200 km/sec, are likely to be the result of gravitational perturbations whose characteristic dimension is that of a cluster of galaxies (that is, $\sim 1 \times 10^6$ pc). The presence of these random velocities makes it impossible to use galaxies less distant than $\sim 3 \times 10^6$ pc to determine the Hubble constant accurately.

Because of the relative faintness of Cepheids, bright supergiants, large HII regions, and ordinary novae must be used as distance indicators to galaxies more distant than $\sim 3 \times 10^5$ pc. These latter distance indicators cannot be used for galaxies more distant than $\sim 2 \times 10^7$ pc, and thus the Hubble constant can only be determined by means of galaxies more distant than $\sim 3 \times 10^6$ pc, but less distant than $\sim 2 \times 10^7$ pc.

Before bright supergiants, large HII regions, and ordinary novae can be used as distance indicators, they must be calibrated in relatively nearby galaxies where Cepheid variables, which are more precise distance indicators, can be accurately observed. Galaxies within the local group, the M81 group, and the South Polar group are the most important galaxies for this calibration.

In principle, the cosmological parameters H and q can be determined from the observable galaxy magnitude-redshift relation. It can be shown that all Friedmann models with $\Lambda = 0$ [see Equations (13–29) and (13–30)] and $q > 0$ obey a rigorous relation between apparent bolometric magnitude and redshift.

$$m_{bol} = 5 \log \left[\frac{qz + (q - 1)(\sqrt{1 + 2qz} - 1)}{q^2} \right] + \text{constant.} \qquad (13\text{--}1)$$

In the limit of small z, we find

$$m_{bol} = 5 \log z + 1.086(1 - q)z + \text{constant} \qquad (13\text{--}2)$$

for all q. The leading term in the above equations, which implies that galaxies are receding, has been confirmed by observations. However, in order to determine q, it is necessary to measure the magnitude-redshift relation for galaxies with relatively large redshift (that is, $z \gtrsim 0.4$). Since $z \sim HD/c$ with $H \sim 75$ km/sec/Mpc*, the above requirement means that galaxies more distant than $D \sim 2 \times 10^9$ pc must be used to determine q. The brightest galaxies in a rich cluster of galaxies are the only available distance indicators at such great distances.

Giant elliptical galaxies are the brightest galaxies in rich clusters. The Virgo and Coma clusters contain giant elliptical galaxies such as M87, as well as spiral galaxies, which in turn contain bright supergiants and HII regions. These two clusters are used to determine the distances to bright elliptical galaxies. This is believed necessary because supernovae and globular clusters, which are the brightest resolved objects in elliptical galaxies, are not generally regarded as reliable distance indicators. However, attempts have recently been made to estimate the distance to M87 by means of a comparison between the luminosity functions of globular clusters in M87 and those in the nearby giant spiral galaxy M31 (Andromeda). This method is quite uncertain, since it is not clear that the luminosity functions for globular clusters should be similar in two different galaxies. The most recent observational data for more than 40 clusters of galaxies with the brightest cluster member as distance indicator gives the result $q \sim 1$. However, this result is very uncertain and probably too large.

Several corrections must be applied before galaxies can be used as distance indicators. The first correction is necessary because galaxies do not have well-defined boundaries, and therefore a standard diameter must be defined (see Section 12–1). A second correction is necessary, because the photomultipliers used in the measurements have fixed-frequency bandwidths. On the other hand, different parts of the spectrum are measured from galaxies with different z. This correction is called the K effect. Because the giant elliptical galaxies have similar spectra, a universal K correction can be

---

* Recent determinations of the Hubble constant indicate that $H \sim 50$ km/sec/ Mpc.

made if evolutionary effects are small. The possibility of significant inter-galactic absorption introduces a further uncertainty.

Since it is likely that galaxy formation took place during a single epoch, evolutionary effects are expected to systematically change the luminosity and spectra of galaxies as a function of world time. The metal content and consequently the opacities of stars in elliptical galaxies are higher than those of population II stars. This circumstance implies that stars in elliptical galaxies, which are as old as population II stars, are more massive (see Section 2–8) and also have redder giant branches. They resemble old disk stars in our own galaxy. Although observations show that evolution of color is small, recent estimates indicate that the luminosity may evolve significantly for elliptical galaxies even with $z \leq 0.5$.

The use of the brightest galaxy in a rich cluster as a distance indicator assumes that there is a well-defined upper limit to the brightness of galaxies in clusters. Recent observations have suggested that the dispersion in the absolute magnitude of the brightest galaxy in a rich cluster is ~ 0.3 magnitudes. If no such upper limit to the luminosity of giant ellipticals existed, then we would systematically find intrinsically brighter galaxies in very distant clusters and thereby obtain a value of q from the magnitude-redshift relation that is too high. The size of this systematic error would depend critically on the poorly determined slope of the cluster luminosity function.

A method of overcoming this difficulty has been suggested. Because radio and optical brightnesses are not correlated, a magnitude-redshift relation based on a complete sample of radio galaxies would avoid this difficulty. Since approximately 30 percent of radio galaxies are in rich clusters, they could be used to identify very distant clusters of galaxies that would otherwise be too faint to be readily discovered.

13–2   METRIC TENSOR AND LINE ELEMENT

If two events in space time are labeled by coordinates x^i and $x^i + dx^i$ ($i = 0, 1, 2, 3$), the square of the proper distance between them is

$$ds^2 = g_{ij}\, dx^i\, dx^j. \tag{13-3}$$

For flat space, a coordinate system may be found whose metric tensor is

$$g_{ij} = \begin{pmatrix} 1 & 0 & 0 & 0 \\ 0 & -1 & 0 & 0 \\ 0 & 0 & -1 & 0 \\ 0 & 0 & 0 & -1 \end{pmatrix}. \tag{13-4}$$

When we are dealing with very large regions of space, the above expression for g_{ij} is inadequate, since space has curvature.

If ds^2, which is called the line element, is negative, the two events are separated by a spacelike interval, and $(-ds^2)^{1/2}$ is the distance between two events as measured by an observer who sees these events simultaneously. On the other hand, if ds^2 is positive, the two events are separated by a time-like interval, which implies that an observer can travel from one event to the other. For timelike intervals, ds is $c \times$ the proper time interval that would be measured by an observer who moves from one event to the other. The path of a light signal or any signal traveling at the limiting velocity satisfies the condition $ds^2 = 0$.

13-3 PRINCIPLE OF EQUIVALENCE

The principle of equivalence, which is the basis for the theory of general relativity, asserts that even in curved space, an observer can choose a locally Euclidean reference frame with himself at the center. This principle implies that general relativity is equivalent to special relativity if only small regions of space are considered, and consequently the laws of physics as applied to a small region of space are the same in curved space as they are in flat space.

13-4 COSMOLOGICAL PRINCIPLE AND CO-MOVING OBSERVERS

Most cosmological theories are based on the cosmological principle, which asserts that the universe is homogeneous and isotropic.

A co-moving observer is defined to be an observer who is always at rest with respect to matter in his vicinity. Along the path of a co-moving observer, the metric tensor can always be chosen to be locally of the form given in Equation (13–4) with the first space derivatives of g_{ij} equal to zero at the position of the observer. The cosmological principle implies that the syn-chronization of time can be made by instructing each co-moving observer to use the density in his neighborhood as a measure of time. Therefore, we can define a world time t, which is common to all co-moving observers.

13-5 ROBERTSON-WALKER LINE ELEMENT

If we assume that the universe is homogeneous and isotropic (the cosmolog-ical principle), it can be shown that the most general expression for the line element (called the Robertson-Walker line element) is

$$+ds^2 = c^2 \, dt^2 - R^2(t) \, du^2 \tag{13-5}$$

where du is defined in a 3-space of constant curvature. We can write du in the form

$$du^2 = \frac{dr^2}{(1 - kr^2)} + r^2(d\theta^2 + \sin^2 \theta \, d\phi^2) \tag{13-6}$$

where r, θ, and ϕ are co-moving coordinates, and k, the curvature index, can be either 1, 0, or -1. The space is said to be closed if $k = 1$, open if

$k = -1$, and flat if $k = 0$. If space is closed, the surface area of a sphere of radius r is less than $4\pi r^2$. If space is open, the surface area of a sphere of radius r is greater than $4\pi r^2$. The Robertson-Walker line element implies a kinematic model of the universe that is unique except for the arbitrary expansion function $R(t)$ and the curvature index k.

13–6   REDSHIFT

The Robertson-Walker line element implies that in a homogeneous and isotropic universe, the distance between two co-moving observers can be written

$$D(t) = D_0 R(t) \tag{13-7}$$

where D_0 is a constant that depends on the coordinate distance between the two observers, and $R(t)$, the expansion parameter, is a universal function of the world time. The following simple model will help clarify Equation (13–7). A sphere is an obvious example of a closed surface with constant curvature. Imagine an expanding sphere with observers located at fixed points on its surface. It is clear that although the distance between observers changes as a function of world time, the relative distance does not change, and consequently Equation (13–7) is valid.

It follows from Equation (13–7) that the relative velocity of two co-moving observers is

$$V = \frac{dD(t)}{dt} = D_0 \frac{dR(t)}{dt}. \tag{13-8}$$

The Hubble constant, which is a function of world time except in the case of the steady-state cosmology, is defined by the relation

$$H(t) \equiv \frac{1}{R(t)} \frac{dR(t)}{dt}. \tag{13-9}$$

Equations (13–8) and (13–9) imply the Hubble law

$$V = H(t)D(t). \tag{13-10}$$

The Hubble constant is determined from the observed redshifts of galaxies. Therefore, it is important to show how the expansion parameter and Hubble constant are related to the observable redshift. In Section 13–2 we noted that the path of a light signal satisfies the relation $ds^2 = 0$. We consider a light signal of frequency v_0 emitted from 0 at time t_0 and received by a co-moving observer P at time t_1. The coordinate distance between 0 and P is u. It follows from Equation (13–5) that the relation between coordinate distance and the expansion parameter $R(t)$ is

$$u = c \int_{t_0}^{t_1} \frac{dt}{R(t)}. \tag{13-11}$$

Equation (13–11) shows that if a light signal is emitted at 0 during the world time interval $t_0 \leq t \leq t_0 + dt_0$ and received at P during the corresponding interval $t_1 \leq t \leq t_1 + dt_1$, then dt_0 and dt_1 satisfy the equation

$$\frac{dt_0}{R(t_0)} = \frac{dt_1}{R(t_1)}.$$ (13–12)

Since the number of emitted wavelengths must be the same as the number of received wavelengths, v_0 and v_1 must be related by the expression

$$v_0 \, dt_0 = v_1 \, dt_1.$$ (13–13)

Equations (13–12) and (13–13) imply

$$\frac{v_0}{v_1} = \frac{\lambda_1}{\lambda_0} = \frac{R(t_1)}{R(t_0)}$$ (13–14)

where λ_0 and λ_1 are the wavelengths of the emitted and received light signals, respectively.

Equation (13–14) shows that the observed redshift $z = (\lambda_1 - \lambda_0)/\lambda_0$ is related to the expansion parameter by the relation

$$1 + z = \frac{R(t_1)}{R(t_0)}$$ (13–15)

which to first order in $t_1 - t_0$ becomes

$$1 + z \sim 1 + \frac{1}{R(t_0)} \frac{dR(t_0)}{dt} (t_1 - t_0)$$ (13–16)

or

$$z \sim H(t_0)(t_1 - t_0) \sim H(t_0) \frac{D}{c}$$

where D is the distance between two neighboring co-moving observers. Equations (13–8) and (13–16) show that to first order in $t_1 - t_0$, we have $v \sim cz$, and therefore for small redshifts, it is correct to interpret the observed redshift z as a Doppler shift.

13–7 TEMPERATURE AND SURFACE BRIGHTNESS OF RADIATION

We assume that at some world time t_0, the universe is filled with blackbody radiation of temperature T_0, which is entirely decoupled from matter. We wish to determine how this radiation evolves as a function of world time. For blackbody radiation, the number of photons at frequency v_0 per unit volume in the bandwidth dv_0 is

$$dn_0 = \frac{8\pi v_0{}^2}{c^3} \frac{dv_0}{[\exp(hv_0/kT_0) - 1]}$$ (13–17)

at time t_0. Therefore, the number of photons in a volume element V_0 at t_0 is $V_0 \, dn_0$. Since the total number of photons is conserved during expansion, the assumed symmetry of the universe implies that the number of photons in any volume element such as V_0 should also be conserved during expansion. It is also obvious that as the universe expands, the volume of the element V_0 becomes

$$V_1 = V_0 \left[\frac{R(t_1)}{R(t_0)} \right]^3. \tag{13-18}$$

It follows from Equation (13-14) that photons of frequency v_0 and bandwidths dv_0 at time t_0 are shifted to frequency

$$v_1 = v_0 \frac{R(t_0)}{R(t_1)} \tag{13-19}$$

and bandwidth

$$dv_1 = \frac{R(t_0)}{R(t_1)} dv_0 \tag{13-20}$$

at world time t_1. It can readily be shown by means of Equations (13-17) through (13-20) and the condition $V_0 \, dn_0 = V_1 \, dn_1$ that the temperature at world time t_1 is

$$T_1 = T_0 \frac{R(t_0)}{R(t_1)}. \tag{13-21}$$

Equation (13-21) shows that as the universe expands the temperature varies inversely as the expansion parameter. Since the surface brightness B of the blackbody radiation varies as T^4, we find

$$\frac{B(t_1)}{B(t_0)} = \left[\frac{R(t_0)}{R(t_1)} \right]^4. \tag{13-22}$$

Although the above expression for B is for blackbody radiation, it is clear that it is general, since the path of a given photon does not depend on the shape of the spectrum.

13-8   ANGULAR SIZE AND APPARENT MAGNITUDE

The path of a light signal satisfies the null geodesic $ds^2 = 0$ in curved as well as flat space. Therefore, it follows from Equation (13-6) that if an observer is at $r = 0$ along the light path, we have

$$\int_0^{r_1} \frac{dr}{\sqrt{1 - kr^2}} = u = \int_{t_0}^{t_1} \frac{c \, dt}{R(t)} \tag{13-23}$$

where t_0 and t_1 are the times of emission and reception, respectively, of the light signal emitted at coordinate position r_1. The left-hand side of Equation (13–23) integrates to

$$\begin{aligned} u &= \sin^{-1} r_1 && \text{if} \quad k = +1 \\ u &= r_1 && \text{if} \quad k = 0 \\ u &= \sinh^{-1} r_1 && \text{if} \quad k = -1. \end{aligned} \qquad (13\text{–}24)$$

The metric distance at the time of photon emission is $R(t_0)r_1$. It follows from Equation (13–24) that $r_1 = \sin u$, u, or $\sin hu$ for $k = 1, 0$, or -1, respectively. The observed angular diameter of linear dimension l, whose metric distance is $R(t_0)r_1$ at the time of photon emission, is

$$\theta = \frac{l}{R(t_0)r_1} = \frac{l(1 + z)}{R(t_1)r_1} \qquad (\theta \ll 1) \qquad (13\text{–}25)$$

where the second relation follows from Equation (13–15). It is readily seen that Equation (13–25) reduces to the usual definition of angular diameter in flat space when $u = r_1$.

Let L be the luminosity of a galaxy at world time t_0 and F the observed power per unit area at world time t_1. Recalling that the brightness is the energy flux per unit solid angle and that F must reduce to the classical result

$$F = \frac{L}{4\pi c^2(t_1 - t_0)^2} \qquad (13\text{–}26)$$

at small distance, it follows from Equations (13–22) and (13–25) that

$$F = \theta^2 B(t_1) = \frac{L}{4\pi R^2(t_0)r_1{}^2} \left[\frac{R(t_0)}{R(t_1)}\right]^4 \qquad (13\text{–}27)$$

which can obviously be rewritten as

$$F = \frac{L}{4\pi R^2(t_1)r_1{}^2} \left[\frac{R(t_0)}{R(t_1)}\right]^2. \qquad (13\text{–}28)$$

Equation (13–28) shows that the energy flux is the luminosity divided by the area of the sphere centered on the observer at t_1, the time of reception multiplied by $[R(t_0)/R(t_1)]^2$, which by Equation (13–28) is seen to be the Doppler shift squared. One factor of $[R(t_0)/R(t_1)]$ results because the recession of the co-moving observers increases the relative time of measurement of the light at the observer as compared with the corresponding time of emission. The energy of a photon is not a conserved quantity, since it is the time component of a four vector. The second Doppler factor in Equation (13–28) takes into account this change in energy as the photon travels from source to observer.

13–9 FUNDAMENTAL EQUATIONS OF COSMOLOGY

If the Robertson-Walker line element [see Equation (13–5)] is assumed to hold, it can be shown that the Einstein field equations reduce to the following equations for $R(t)$:

$$\frac{\dot{R}^2}{R^2} + 2\frac{\ddot{R}}{R} + \frac{8\pi GP}{c^2} = -\frac{kc^2}{R^2} + \Lambda c^2 \tag{13–29}$$

$$\frac{\dot{R}^2}{R^2} - \frac{8\pi G\rho}{3} = -\frac{kc^2}{R^2} + \frac{\Lambda c^2}{3} \tag{13–30}$$

where $\dot{R} \equiv dR(t)/dt$, and Λ is the cosmological constant, which is generally taken to be zero. If Equation (13–30) is subtracted from Equation (13–29) and Λ set equal to zero, we find

$$\frac{\ddot{R}}{R} + 4\pi G\left(\frac{\rho}{3} + \frac{P}{c^2}\right) = 0. \tag{13–31}$$

The definitions of H and q given in Equation (13–9) and Equation (1–8) imply that Equation (13–31) can be reduced to

$$\frac{3H^2 q}{4\pi G} = \rho + \frac{3P}{c^2}. \tag{13–32}$$

If Equation (13–30) is substituted into Equation (13–32), we find

$$\frac{kc^2}{R^2} = \frac{4\pi G}{3q}\left[\rho(2q - 1) - \frac{3P}{c^2}\right]. \tag{13–33}$$

Equations (13–32) and (13–33) relate the curvature of space at a particular epoch with the energy content of the universe as indicated by its total density and pressure.

At the present epoch, the pressure P is negligible, and consequently Equations (13–32) and (13–33) reduce to

$$\rho_0 = \frac{3H_0^2 q_0}{4\pi G} = 2 \times 10^{-29} q_0 \text{ g/cm}^3 \tag{13–34}$$

$$\frac{kc^2}{R_0^2} = H_0^2(2q_0 - 1) \tag{13–35}$$

if H_0 is taken equal to 75 km/sec/Mpc. Equations (13–32) and (13–35) show that the universe will be closed if $q \geq \frac{1}{2}$, which is equivalent to the requirement $\rho_0 \gtrsim 1 \times 10^{-29}$ g/cm^3.

Although negligible at the present epoch, the energy density due to radiation, $E_r = aT^4$, will be much greater than the matter density $\rho_m c^2$ during the early history of the universe, since $aT^4/\rho_m c^2$ varies as $R(t)^{-1}$. If radiation is the dominant source of energy density and pressure, we have $\rho = E_r/c^2$ and $P = E_r/3$, which together with Equation (13–32) imply

$$E_r = \frac{3H^2qc^2}{8\pi G}.$$ (13–36)

From Equations (13–30) and (13–33) we find

$$\frac{kc^2}{R^2} = H^2(q - 1)$$ (13–37)

which shows that a radiation-dominated universe will be closed if $q \geq 1$.

13–10   RADIO COSMOLOGY

The identification of the very intense radio source Cygnus A with a distant ($z = 0.06$) galaxy indicated that the radio properties of very distant galaxies could be measured with high precision and consequently be of potential value to cosmology. In principle, the observation of objects at very large distances ($z \gtrsim 1$) could discriminate among cosmological models. The great potential advantage of radio as compared with optical observations of galaxies is illustrated by the radio galaxy 3C295 which has a redshift of $z = 0.46$. Although galaxies with such a large redshift are very faint when observed in the visual region of the spectrum, 3C295 is one of the strongest extragalactic radio sources. Unfortunately, the very large intrinsic dispersion in the physical properties (that is, energy flux, polarization, spectrum, and linear dimension) of radio galaxies limits their application to cosmology. Only radio counts, which determine the number of radio sources down to some limiting energy flux F_v, have yielded results that are of cosmological interest.

If all radio sources were uniformly distributed throughout Euclidean space with the same luminosity, then the number of sources inside a volume of radius R would be proportional to R^3. Since the energy flux F_v decreases as $1/R^2$ in Euclidean space, the number of sources N with energy flux greater than F_v would satisfy the relation

$$\log N = -\tfrac{3}{2} \log F_v + K$$ (13–38)

where K is a constant. The $-\tfrac{3}{2}$ slope of the $\log N$–$\log F_v$ relation given in Equation (13–38) is characteristic of Euclidean space. Most cosmological models predict a greater slope (~ -1.2).

The results of radio counts are summarized in Figure 13–1. These results indicate that the number of sources in different flux ranges is not consistent with static Euclidean space, the steady-state model, or simple Friedmann models. An examination of Figure 13–1 shows that the slope of the $\log N$–$\log F_v$ relation at 408 MHz is ~ -1.5 for relatively high limiting flux levels,

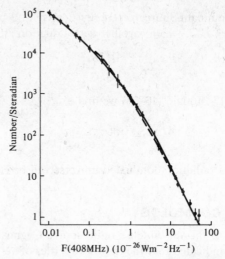

Figure 13–1 The log N-log $F_{(v)}$ relation for $F(408$ MHz$) \gtrsim 0.01 \times 10^{26}$ Wm^{-2}Hz^{-1}. The dashed line corresponds to the 178 MHz source counts scaled by an amount corresponding to a mean spectral index $\alpha = 0.7$.

Source: From Ryle, M. 1968, *Ann. Rev. Astron. Ap.*, **6**, 249, by permission of Annual Reviews, Inc.

decreases to ~ -1.85 until $F_v \leq 4 \times 10^{-26}$Wm^{-2}Hz^{-1}, and then increases to ~ -0.8 for the lowest measured energy flux. The excess of sources in the range of one flux unit[1] and the deficiency of sources at very low flux levels indicate that the mean properties of radio sources are a function of epoch. This result is consistent with "big bang" cosmologies that are suggested by the presence of a 3°K background radiation. However, this interpretation of the radio source counts is by no means certain.

QSOs represent ~ 20–25 percent of radio sources more luminous than F_v ($v = 178$ MHz) $>$ one flux unit. It has been suggested that the observed excess of weak sources is caused by the presence of a large number of relatively faint QSOs. If this suggestion were correct, it would cast doubt on the validity of using radio counts in cosmology, since the cosmological origin of the QSO redshifts is still in doubt.

If log N–log F_v curves are plotted separately for galaxies, QSOs, and unidentified sources, they are found to have slopes of -1.5, -1.8, and -2.3, respectively. If the unidentified sources were primarily faint galaxies, then the slope of the combined galaxy curve would be similar to that of QSOs, a circumstance that would be consistent with their occupying the same volume of space.

---

[1] One flux unit $= 10^{-26}$Wm^{-2}Hz^{-1}.

If most faint radio sources are very distant, then an explanation of the radio counts suggests either that the number of sources and/or the intrinsic power of a typical source was greater at an earlier epoch. The typical radio luminosity must vary as t^{-n} with $n \sim -2.2$, or the mean density vary as t^{-n} with $n \sim -3.5$ in order to explain the radio counts. The apparent cutoff in the observed number of radio sources at flux levels less than one flux unit suggests a reduction in source evolutionary time for epochs corresponding to $z \gtrsim 3$–4. The inverse Compton scattering of relativistic electrons with the background radiation, whose energy density is amplified by a factor of $(1 + z)^4$ at earlier epochs, could reduce the evolutionary time scales of radio galaxies.

13–11 3°K BACKGROUND RADIATION AND THE PRIMEVAL FIREBALL

Cosmological theories attempt to give an overview of the history of the universe. Such theories are of necessity based on a small number of observations, and therefore all evidence that is of cosmological relevance is of great significance. The widespread current interest in cosmology is due, in large measure, to a few recent astronomical measurements. The most notable of these is the discovery of the 3°K background radiation, which is generally interpreted as relic radiation from the primeval fireball. If such an interpretation of the background radiation is accepted, then the thermal history of the universe can be predicted. Although the universal energy content of the 3°K background radiation exceeds that caused by all other photon sources, it is difficult to detect and therefore was not discovered until 1965.

Figure 13–2 shows an antenna, which is placed inside a total absorbing enclosure of uniform temperature T and connected to a resistor held at similar temperature. Thermal fluctuations of the electrons inside the resistor will cause noise voltages (called Johnson noise) to be induced across it. The noise power generated by the resistor in the bandwidth Δv is

$$2kT \, \Delta v \qquad\qquad (13\text{–}39)$$

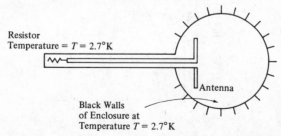

Resistor
Temperature $= T = 2.7°$K

Antenna

Black Walls
of Enclosure at
Temperature $T = 2.7°$K

Figure 13–2 A resistor matched to antenna is in equilibrium with surrounding blackbody.

if both electric field polarizations are transmitted. If the antenna and resistor are properly matched, all the noise power produced by the resistor will be emitted into free space by the antenna.

The power per unit area in the frequency band Δv incident on the antenna is

$$\frac{8\pi k T}{\lambda^2} \Delta v \tag{13-40}$$

if the frequency of the observed radiation is sufficiently low for the Rayleigh-Jeans limit to be valid (that is, $hv \ll kT$). In thermodynamic equilibrium, the power emitted by the antenna must equal that absorbed. It follows that the effective area of the antenna at wavelength $\lambda = c/v$ must equal $\lambda^2/4\pi$ in order that the power absorbed per unit frequency be independent of frequency as is the power emitted by the antenna. Radio measurements reveal the existence of a background radiation whose spectrum is that of a 2.7°K blackbody.

Although the prediction of a blackbody residual radiation caused by a primeval fireball was suggested as early as 1946, the 3°K background radiation was not discovered until 1965. Figure 13–3 shows the spectra of the 3°K background radiation, discrete radio sources and stars. At long wavelengths (≥ 8 mm), the observed spectrum agrees very well with that predicted for a cosmic blackbody background. Moreover, indirect infrared background measurements that are based on measuring the relative equivalent widths of lines from CN (and CH) are consistent with a blackbody spectrum extending into the infrared.

The CN molecule, which is common in interstellar space, is convenient as an indicator of background radiation, because its lowest-lying rotational excited state can be populated by photons close to the 3°K blackbody peak

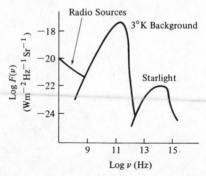

Figure 13–3   The flux density from discrete background radio sources, 3°K background radiation and starlight.

(~ 2.6 mm). Figure 13–4 shows the relevant energy levels of the CN molecule. Absorption lines are produced when the CN molecule absorbs light from a hot star. The relative equivalent widths of the absorption lines can be used to determine the temperature of the radiation field if the relative populations of the energy levels are a result of the background radiation.

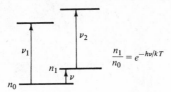

Figure 13–4   The lowest energy levels of CN molecule.

It is important to ask whether other types of radiation besides primordial radiation might explain the observations. Known sources of background radiation such as atmospheric radiation, stellar emission, dust, or observed radio sources are inadequate to explain the observed background. Because radio and infrared photons have very long mean free paths, the observed isotropy of the background radiation demonstrates that whatever the cause of the radiation, its presence is not local. Its energy density (0.5 eV/cm^3) argues that it is of cosmological importance regardless of origin. Arbitrary distributions of very distant sources can be constructed that will mimic the 3°K background radiation. However, the observed isotropy of the background radiation requires that these undefined sources be more numerous than known galaxies.

Measurements of the isotropy of the background radiation provide information about the universe during an early epoch. If the observed density of matter is comparable to the actual density, then the optical depth for scattering by means of free electrons is unity at the epoch of hydrogen recombination (that is, when $T \sim 3500$°K and $z \sim 1000$). Higher mean densities of matter could reduce the typical distance to the last scattering to $z \sim 7$, a distance which is still greater than the most distant known galaxy or QSO. Because electron scattering smooths radiation over characteristic dimensions comparable with the size $\sim c/H$ of the visible universe at the epoch of the last scattering, irregularities that occur during the earliest stages of the universe are not directly observable. However, if the temperature in the early universe were irregular over a characteristic distance smaller than the smoothing provided by electron scattering, the presently observed 3°K background radiation would be a superposition of blackbody spectra, since the energy of a photon is changed only slightly in each scattering.

13–12   THERMAL AND PARTICLE HISTORY OF THE UNIVERSE

The assumption that the 3°K background radiation is of primordial origin implies a thermal history of the universe that is shown in Figure 13–5. The temperature and densities at early epochs are obtained from the corresponding present values T_0 and ρ_0 by means of the transformations

$$T_1 = T_0 \frac{R(t_0)}{R(t_1)}$$

and

$$\rho_1 = \rho_0 \left(\frac{R(t_0)}{R(t_1)}\right)^3 \tag{13–41}$$

so long as matter and radiation are not strongly coupled.

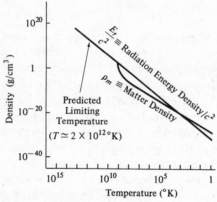

Figure 13–5   Radiation density and matter density as function of temperature in an expanding hot universe.

At the present epoch E_m, the mean energy density in observable matter, dominates the radiation energy density E_r caused by blackbody radiation; that is,

$$\frac{E_r}{E_m} = \frac{aT_0^4}{\rho_0 c^2} \ll 1 \tag{13–42}$$

where T_0 is 2.7°K, and the observed mean matter density is 5×10^{-31} g/cm^3. However, since the radiation energy density varies as R^{-4} as compared to R^{-3} for the density of matter, the radiation energy density will increase faster than that of matter as we go back to earlier epochs. The condition $E_m = E_r$ will be satisfied when

$$\rho_0 c^2 \left(\frac{R(t_0)}{R(t_1)}\right)^3 = aT_0^4 \left(\frac{R(t_0)}{R(t_1)}\right)^4 \tag{13–43}$$

that is, when the temperature is

$$T = T_0 \frac{E_m}{E_r} \sim 10^{2-3\circ}\text{K}. \tag{13-44}$$

The total energy density during the epoch in which matter and radiation energy densities are equal is

$$\frac{2aT^4}{c^2} = \frac{2E_m{}^4}{c^2 E_r{}^3} \sim 10^{-21} \text{ g/cm}^3. \tag{13-45}$$

The primeval fireball (big bang) origin for the 3°K background radiation implies that the temperature and density were very high during the early universe. The world time (that is, epoch) and density are related by the expression

$$t_{ff} = \alpha \sqrt{\frac{3\pi}{32G\rho}}. \tag{13-46}$$

Equation (13-46) is similar to that for the Newtonian free-fall time [see Equation (3-60)] except that α, which is of order unity, depends on the particular cosmological model.

It should be pointed out that quantum fluctuations of the metric tensor set a limit to the earliest world time at which classical cosmology is applicable. It is impossible to speak of events in a well-ordered time sequence for world times less than

$$t \sim \left(\frac{G\hbar}{c^5}\right)^{1/2} \sim 10^{-44} \text{ sec} \tag{13-47}$$

which from Equation (13-46) corresponds to a density of approximately

$$\rho \sim \frac{c^5}{G^2\hbar} \sim 10^{94} \text{ g/cm}^3. \tag{13-48}$$

When the temperature is very high (that is, $T \gtrsim m_\pi c^2/k \sim 2 \times 10^{12\circ}\text{K}$), the universe consists of hadrons (which are strongly interacting baryons and mesons), leptons (for example, ν, ν^-, e^+, e^-, μ^-, μ^+), photons, gravitons, and perhaps quarks. Baryons, which are subclassed as nucleons (p, n), and hyperons (that is, Λ, Σ, Ξ, Ω, and so on) are likely to be the major constituents of the very early universe. Recent theoretical work in particle physics has suggested that the temperature cannot exceed about $2 \times 10^{12\circ}\text{K}$. If this theory is correct, then during the very early history of the universe, matter existed in the form of massive, strange particles whose kinetic energies were nonrelativistic.

The prediction of a limiting temperature is based on empirical evidence concerning the number of different species of elementary particles as a function of energy. The measured density of mass states appears to be increasing exponentially as a function of energy, that is, $\rho(E) \propto e^{bE}$ with $b^{-1} = 160$ MeV. Under conditions of thermodynamic equilibrium, the average energy of a mass state is

$$\bar{E} = \frac{\int E\rho(E)e^{-E/kT}\,dE}{\int \rho(E)e^{-E/kT}\,dE}. \tag{13-49}$$

If $\rho(E)$ increases exponentially with E, then the above integrals will not converge if T is greater than T_{max}, where $kT_{max} \simeq 160$ MeV. This result implies that as the temperature is raised to its limiting value, further increases in energy go into creating new particles rather than increasing the kinetic energy of existing particles.

As the temperature falls below $10^{12\circ}$K, high-energy baryon states are depopulated, and only nucleons remain. This is the case because the expansion time is long as compared to the time for depopulation. When the mean photon energy becomes less than ~ 1 GeV, nucleons annihilate each other; that is,

$$N + \bar{N} \rightarrow \gamma + \gamma$$

and consequently the number of baryons is greatly reduced. The number of baryons minus the number of antibaryons is a conserved quantity called the baryon number. Because the amount of matter and antimatter is almost equal during the earliest stages of the universe, a small fractional difference between the number of particles and antiparticles is of great importance in determining the future properties of the universe. If the number of particles and antiparticles were exactly equal, the universe would become solely radiation, neutrinos, and gravitons at sufficiently low temperature unless the spatial distribution of matter and antimatter were nonuniform.

When the temperature of the universe is between $T = m_e c^2/k \sim 10^{10\circ}$K and $T = m_\pi c^2/k \sim 2 \times 10^{12\circ}$K, the energy content of the universe is primarily in the form of leptons (that is, electrons, positrons, neutrinos, antineutrinos, and muons). This era extends from about world time $\sim 10^{-4}$ sec to ~ 10 sec. As the density and temperature of the universe decrease, muon neutrinos (ν_μ, $\bar{\nu}_\mu$), which initially interact with muons and hadrons, become decoupled from the rest of the universe. Likewise, electron neutrinos (ν_e, $\bar{\nu}_e$), which are initially coupled to electrons and positrons through such interactions as

$$e^- + e^+ \rightleftarrows \bar{\nu}_e + \nu_e$$

and to nucleons by means of the interactions

$$p + \bar{\nu} \rightleftarrows n + e^+$$
$$p + e^- \rightleftarrows n + \nu_e$$
$$n \rightleftarrows p + e^- + \bar{\nu}_e$$

likewise decouple from other particles. Because of the relatively long half-life of the free neutron ($\simeq 10$ min), the first two reactions are dominant so long as the energies of the electrons exceed the n-p mass difference. After the temperature decreases below $10^{10\circ}$K, electron-positron pairs annihilate and thereby produce photons.

At temperatures in excess of about $10^{10\circ}$K, pair formation

$$\gamma + \gamma \rightleftarrows e^+ + e^-$$

takes place. It follows from Equations (2–31), (2–35), and (8–13) that in the relativistic limit when $E = pc$, the energy density and number density of electrons plus positrons are

$$\mathcal{E} = \frac{g}{\pi^2} \frac{(kT)^4}{(\hbar c)^3} \int_0^\infty \frac{x^3 \, dx}{e^x + 1}$$

$$= \tfrac{7}{4} a T^4 \tag{13–50}$$

and

$$n = \frac{g}{\pi^2} \left(\frac{kT}{\hbar c}\right)^3 \int_0^\infty \frac{x^2 \, dx}{(e^x + 1)} \tag{13–51}$$

where g, the degeneracy factor, is 2. Equation (13–50) shows that the inter-particle distance is approximately

$$\frac{\hbar c}{kT} \, .$$

Consequently, the electrostatic energy of interaction is approximately

$$e^2 \frac{kT}{\hbar c} = \alpha kT \ll kT \tag{13–52}$$

and therefore the particles behave as a free gas. α is the fine structure constant.

Neutrinos decouple from matter and radiation when the temperature falls below $10^{10\circ}$K. From this epoch onward, the neutrino temperature varies inversely as the expansion parameter (that is, $T \propto R(t)^{-1}$). The entropy of radiation in a co-moving volume element $V(t)$ is

$$S = \tfrac{4}{3} a T^3 V. \tag{13–53}$$

Since the total entropy is conserved during pair annihilation, we have

$$(1 + \tfrac{7}{4}) \tfrac{4}{3} a T^3 V = \tfrac{4}{3} a T'^3 V \tag{13–54}$$

where the left-hand side of Equation (13–54) denotes the entropy due to electron-positron pairs and radiation before pair annihilation, and the right-hand side denotes the entropy due to radiation after pair annihilation.

Equation (13–54) shows that the photon temperature T' is related to the neutrino temperature T by the relation

$$T' = \left(\frac{11}{4}\right)^{1/3} T. \qquad (13\text{–}55)$$

Equation (13–55) follows from Equation (13–54) because the neutrino temperature equals the radiation temperature prior to pair annihilation.

The production of helium commences after annihilation processes have ceased. If the early expansion of the universe can be described by means of simple Friedmann cosmological models, the amount of helium produced is determined by the n/p ratio at the world time helium production commences (~ 30 sec) and the relevant nuclear reaction cross sections. The principal nuclear reactions that govern the production of helium are

$$n + p \rightarrow D^2 + \gamma$$
$$D^2 + D^2 \rightarrow He^3 + n$$
$$\rightarrow H^3 + p$$
$$He^3 + n \rightarrow H^3 + p$$
$$D^2 + H^3 \rightarrow He^4 + n.$$

The first of the above reactions, which is an electromagnetic reaction, is the slowest and therefore controls the amount of helium production. For this reason, the n/p ratio at the beginning of helium formation governs the ultimate amount of helium production. Except for the relatively slow decay of free neutrons (~ 10 min), the n/p ratio is nearly frozen in after the first two reactions have ceased. Calculations predict a helium abundance of ~ 30 percent by mass. This result is close to what is presently observed in population I stars. Early anisotropy or the possible relevance of the Brans-Dicke cosmology could lead to an accelerated expansion of the early universe that could reduce the helium production. A degenerate sea of either v_e's or $\bar{v}_e$'s could also change the cosmic helium production by changing the initial n/p ratio.

Except for neutrinos and possibly gravitons, which are decoupled from matter, radiation is the dominant form of energy after $e^- - e^+$ pair annihilation is completed. The radiation density is

$$E_r \sim aT^4 \sim kT \left(\frac{kT}{\hbar c}\right)^3 \qquad (13\text{–}56)$$

where kT is approximately the mean energy per photon, and $(kT/\hbar c)^3$ is approximately the number of photons per unit volume. Equation (13–56) shows that even during the present epoch when matter energy density is much greater than photon energy density, the number density of photons remains greater than the number density of nucleons (by a factor of $\sim 10^9$).

It is convenient to combine Equations (13–41) and (13–46) and find a relation between temperature and world time; that is,

$$T \sim 10^{10}t^{-1/2}°K \qquad (t \text{ small}). \qquad (13\text{–}57)$$

During the early evolution of the universe, hydrogen (and helium) is completely ionized. However, it follows from the Saha equation and the cosmic temperature-density relation that hydrogen will begin to recombine at the epoch when $T \sim 3500°K$ and $\rho \sim 10^{-21}$–10^{-22} g/cm^3. Equation (13–57) shows that this epoch corresponds to a world time of $\sim 10^{13}$ sec. Since the hydrogen recombination cross section is $\sim 10^{-20}$ cm^2 at the relevant temperature, it follows that $\langle \sigma v \rangle \sim 10^{-13}$. For this reason, the mean recombination time, $\tau = 1/(n_e \langle \sigma v \rangle)$, is $\sim 10^{11}$ sec, and it is correct to say that the recombination time is much less than the corresponding world time.

It is important to examine more carefully the validity of the assumption that the background radiation and ionized matter remain thermalized as the universe expands. The primary coupling between matter and radiation is by means of free electron scattering (that is, Thomson scattering). For Thomson scattering, photons with energy less than kT will change frequency by an average amount

$$\frac{\langle \Delta v \rangle}{v} = 4 \frac{\langle v^2/c^2 \rangle}{3} = \frac{4kT}{mc^2}. \qquad (13\text{–}58)$$

Equation (13–58) shows that the maximum rate of energy transfer between ionized matter and radiation is governed by

$$\frac{\dot{E}_r t}{E_r} \sim 4\sigma_T n_e c \left(\frac{kT}{mc^2} \right) t \sim 10^{-10} z^{5/3} \qquad (13\text{–}59)$$

where t is the world time, and the present mean density is assumed to be $\sim 1 \times 10^{-29}$ g/cm^3. The ratio of radiation energy density to thermal energy density is

$$\frac{E_r}{U} = \frac{aT^4}{3n_e kT} \sim 10^8. \qquad (13\text{–}60)$$

This ratio is much greater than unity and nearly independent of epoch. To preserve thermal coupling in an expanding universe, the rate of energy transfer to the plasma must satisfy

$$\frac{\dot{E}_r t}{E_r} > \frac{2}{3} \frac{U}{E_r} \sim 10^{-8}. \qquad (13\text{–}61)$$

The above inequality is satisfied so long as matter remains ionized; consequently, matter and radiation will remain coupled. However, after hydrogen has recombined, the photon mean free path will become greater than the corresponding Hubble distance. Under these conditions, matter will cool faster than radiation unless it is heated by external sources. It is clear that recombination radiation itself can have only a small influence on the background, since the number of recombining atoms is much less than the corresponding number of blackbody photons.

Formation of Galaxies

If the cosmic fireball origin of the background radiation is accepted, then, as will be shown below, radiation drag will inhibit the formation of condensations prior to the recombination of hydrogen. However, after matter and radiation have decoupled, condensations larger than a Jeans length λ_J (see Section 3–3) will be unstable to gravitational collapse. The critical mass for condensation is

$$M_J = \rho\lambda_J^3 \simeq \left(\frac{kT}{Gm_p}\right)^{3/2} \frac{1}{\rho^{1/2}}. \tag{13-62}$$

The value of M_J during the epoch of hydrogen recombination is $\sim 10^6$ M$\odot$, which is approximately the mass of a dwarf galaxy or globular cluster. This circumstance has led to the suggestion that globular clusters may form very early in the history of the universe. The early formation of supermassive stars is another possibility.

We ask how large a perturbation must be if it is to grow prior to hydrogen recombination. The radiation drag per unit volume is

$$F_{\text{drag}} = \frac{v}{c}\, n_e\sigma_T\tfrac{1}{3}E_r \tag{13-63}$$

where σ_T is the Thomson cross section, and v is the relative velocity between the background radiation, which is expanding with the universe, and the matter, which is condensing. It is clear that v must equal or exceed lH, where l is the dimension of the condensation and H the Hubble constant if it is to grow appreciably during the expansion of the universe. On the other hand, the gravitational force per unit volume is

$$F_{\text{grav}} = Gl\rho^2. \tag{13-64}$$

If a condensation is to grow appreciably, it is necessary that

$$F_{\text{grav}} > F_{\text{drag}} = n_e lH\sigma_T \frac{1}{3}\frac{E_r}{c}. \tag{13-65}$$

It follows from Equation (13–65) that radiation drag will inhibit the growth of fluctuation prior to hydrogen recombination.

The above result for the critical mass may appear puzzling, since most of the observed mass is in the form of galaxies of considerably larger mass ($\sim 10^{10}$–10^{12} M$\odot$). However, it is possible that sufficiently massive fluctuations present during the very early stages of the universe might survive the radiation dominated era and subsequently condense. For a perturbation to survive, it is necessary that the thermal diffusion time τ for photons to escape from the perturbation be greater than the Hubble time $1/H$. The characteristic diffusion time τ and dimension l of the perturbation are

$$c\tau \sim \frac{l^2}{\lambda_{\text{photon}}} \qquad (13\text{–}66)$$

where the photon mean free path is $\lambda_{\text{photon}} = 1/n\sigma_T$. The condition $\tau \gg 1/H$ implies

$$\tau H \sim \frac{\Delta E_r}{E_r} = \frac{4\Delta T}{T} \gg 1 \qquad (13\text{–}67)$$

where ΔT and ΔE_r are the differences in thermal energy and temperature, respectively, between the interior of the condensation and the surrounding medium. It follows from expressions (13–66) and (13–67) that the dimension of an optically thick condensation is

$$l \gtrsim \sqrt{\frac{c\lambda_{\text{photon}}}{H} \frac{\Delta T}{T}}. \qquad (13\text{–}68)$$

It follows from Equation (13–68) that only very large and massive ($\gtrsim 10^{12}$ M$\odot$) perturbations can survive the radiation era.

Interaction of Background Radiation with Cosmic-Ray Particles

The number density of blackbody photons is ~ 400 cm^{-3} at the present epoch and increases as $(1 + z)^3$ as we go back to earlier epochs. It can be shown that inverse Compton scattering of the background radiation with cosmic-ray electrons will lead to a reduction in the number of electrons with energies $\gtrsim 10^{10}$ eV. The possible importance of inverse Compton scattering on the evolution of radio galaxies has already been discussed in Chapter 12.

The background radiation will also interact with cosmic γ rays and very-high-energy protons. Pair production caused by high-energy ($\sim 2 \times 10^{14}$ eV) γ rays interacting with background radiation photons would reduce their mean free path through intergalactic matter to less than 30 Mpc. The interaction between γ rays and microwave photons will occur only if there exists an inertial frame of reference such that the energies of both photons exceed the electron rest mass mc^2 so that electron-position pairs can be produced. In the co-moving frame, the microwave photon has an energy of 10^{-3} eV. If the energy of the γ ray relative to the co-moving frame is E, then in the frame such that the energies of the γ ray and microwave photon are the same, we must have

$$\frac{E}{\gamma} = 10^{-3}\gamma \gtrsim mc^2 \text{ eV} \qquad (13\text{–}69)$$

where

$$\gamma = \frac{1}{\sqrt{1 - (v^2/c^2)}}$$

v is the velocity relative to the co-moving frame, and m is the mass of an electron for electron-position pairs to be produced. Solving Equation (13–69), we find then $E \gtrsim 2 \times 10^{14}$ eV, and consequently γ rays more energetic than this amount will interact with microwave photons to form electron-position pairs as they traverse intergalactic space. In addition, protons whose energies exceed 10^{20} eV will be attenuated as a result of photomeson production. For this reason, the high-energy spectrum of protons should decrease sharply if they originate beyond 10 Mpc.

The precise location of high-energy particle cutoffs depends on the high-frequency end of the background radiation. For example, the estimated proton flux and measured upper limits on the γ ray flux from the Crab nebula can be used to obtain upper limits on the energy content of the background radiation at wavelengths shorter than 1.7 mm. These upper limits are consistent with CN and CH measurements.

Diffuse X-Ray Background

X-ray emission ($\frac{1}{2}$ keV–1 MeV) is observed from discrete sources, which are primarily of galactic origin. In addition to discrete x-ray sources, there exists a diffuse x-ray component. The high degree of isotropy of this latter component indicates that it is of extragalactic origin and must consist of at least 10^6 sources. The spectrum of the x-ray background radiation has a nearly continuous spectral index between 1 keV–1 MeV with a change of slope at approximately 40 keV.

Normal galaxies are probably not sufficiently intense x-ray sources to account for the observed background, and therefore it is likely that it originates either from intergalactic space or from nonthermal sources such as Seyfert galaxies, quasistellar objects, and radio galaxies. The inverse Compton effect or nonthermal electron bremsstrahlung are the most likely physical processes responsible for the diffuse x-ray background.

One theory for the diffuse background assumes that it is produced by the interaction of the microwave background radiation with relativistic electrons in intergalactic space. Since the energy of photons produced by means of the inverse Compton effect is

$$\varepsilon \approx \gamma^2 \varepsilon_{3°K} \qquad (13\text{–}70)$$

where $\varepsilon_{3°K}$ ($\sim 3 \times 10^{-3}$ eV) is the energy of the microwave photons and γmc^2 the energy of the relativistic electrons, electrons with $\gamma \sim 5 \times 10^3$ are required if scattering with 3°K background photons is to produce 10-keV x-ray radiation. Presumably the necessary intergalactic electrons are ejected into intergalactic space from quasistellar objects and Seyfert galaxies. Unfortunately, this theory does not explain the flattening of the spectrum at 40 keV. Inverse Compton scattering of relativistic electrons by means of infrared photons in the nuclei of galaxies is another plausible mechanism for producing cosmic x-ray radiation.

Another theory for the origin of the x-ray background assumes that it is bremsstrahlung emission from nonthermal electrons (for example, electrons with a power law energy distribution function). The x-ray spectrum (including the break at 40 keV) can be explained by a suitable combination of continuous ejection and bursts of relativistic electrons from quasistellar objects and radio galaxies at a cosmological epoch corresponding to $z \sim 2$–3.

It is likely that the origin of the x-ray background is related to the origin of cosmic rays (especially very-high-energy cosmic rays) as well as to the microwave background radiation. If either the inverse Compton model or nonthermal bremsstrahlung model is correct, then the intergalactic cosmic-ray flux must be at least 10^{-2}–10^{-4} times the galactic cosmic-ray flux.

APPENDIX A

Einstein Coefficients for Absorption and Emission of Radiation

Consider two stationary energy levels E_l and E_k. The degeneracies of these levels are g_l and g_k, respectively. When a radiative transition takes place between the two levels, a photon of frequency

$$v_{lk} = \frac{E_l - E_k}{h} \tag{A-1}$$

is either emitted or absorbed. The probability that an atom initially in the lower-energy state E_k will absorb a photon of frequency v_{lk} is

$$B_{kl} U(v_{lk})$$

where $U(v_{lk})$ is the energy per unit volume between frequency v_{lk} and $v_{lk} + dv_{lk}$, and B_{kl} is the Einstein coefficient for absorption. The probability that a system initially in the upper state E_l will emit a photon of frequency v_{lk} is the sum of two terms, namely,

$$A_{lk} + B_{lk} U(v_{lk})$$

where A_{lk} is the Einstein coefficient for spontaneous emission, and B_{lk} is the Einstein coefficient of induced emission.

It follows from the principle of detailed balance that under conditions of thermodynamic equilibrium, the number of upward transitions per unit time must equal the number of downward transitions per unit time. For thermodynamic equilibrium, the density of radiation per unit frequency interval is

$$U(v) = \frac{8\pi h v^3}{c^3} \frac{1}{(e^{hv/kT} - 1)}. \tag{A-2}$$

If N_k and N_l are the number of atoms in states k and l, respectively, we have

$$N_k B_{kl} U(v_{lk}) = N_l [A_{lk} + B_{lk} U(v_{lk})]. \tag{A-3}$$

Since N_k and N_l are related by the Boltzmann relation

$$\frac{N_l}{N_k} = \frac{g_l}{g_k} e^{-(h\nu_{lk}/kT)} \tag{A-4}$$

Equation (A–3) reduces to

$$U(\nu_{lk}) = \frac{g_l A_{lk}}{\left[g_k B_{kl} e^{h\nu_{lk}/kT} - g_l B_{lk}\right]} \tag{A-5}$$

Equations (A–2) and (A–5) imply that the Einstein coefficients are related by means of the expressions

$$A_{lk} = \frac{8\pi h \nu_{lk}{}^3}{c^3} B_{lk}$$

$$B_{lk} = \frac{g_k}{g_l} B_{kl}. \tag{A-6}$$

APPENDIX B
Eddington Approximation

The equation of radiative transfer can be written as

$$\frac{dI_\nu(\theta)}{ds} = -k_\nu I_\nu(\theta) + j_\nu \tag{B-1}$$

where θ is the angle between the normal to the surface (taken to be the z axis) and the direction under consideration. I_ν, j_ν, and k_ν are defined in Section 1–6. If the absorption coefficient k_ν is assumed independent of frequency, Equation (B–1) can be rewritten as

$$\cos\theta \frac{dI(\theta)}{d\tau} = -I(\theta) + \frac{j}{k} \tag{B-2}$$

where the optical depth τ of a point that is a distance z beneath the surface is defined by the expression

$$\tau = \int_z^0 k\,dz. \tag{B-3}$$

We define the quantities

$$J = \frac{1}{4\pi} \int I(\theta)\,d\omega$$

$$H = \frac{1}{4\pi} \int I(\theta)\cos\theta\,d\omega \tag{B-4}$$

$$K = \frac{1}{4\pi} \int I(\theta)\cos^2\theta\,d\omega$$

with

$$d\omega = 2\pi \sin\theta\,d\theta.$$

271

Taking moments of Equation (B–2), we find

$$\frac{dH}{d\tau} = J - \frac{j}{k} \tag{B–5}$$

$$\frac{dK}{d\tau} = H. \tag{B–6}$$

If no sources of radiation are present, it follows from Equation (B–5) that

$$j = kJ \tag{B–7}$$

which implies

$$K = H\tau + \text{constant.} \tag{B–8}$$

The angular dependence of the intensity of radiation is assumed to have the simple form

$$\begin{aligned} I &= I_1 \qquad 0 < \theta < \tfrac{1}{2}\pi \\ I &= I_2 \qquad \tfrac{1}{2}\pi < \theta < \pi. \end{aligned} \tag{B–9}$$

Therefore, from Equations (B–4) and (B–9), we find

$$\begin{aligned} J &= \tfrac{1}{2}(I_1 + I_2) \\ H &= \tfrac{1}{4}(I_1 - I_2) \\ K &= \tfrac{1}{3}J. \end{aligned} \tag{B–10}$$

The condition that there be no inflow of radiation at the surface implies that $I_2 = 0$ at $\tau = 0$. Therefore, $J = 2H$ at $\tau = 0$, and it follows from Equations (B–8) and (B–10) that

$$J = H(2 + 3\tau). \tag{B–11}$$

The flux at the surface is defined by the expression

$$F = \int_0^{\frac{1}{2}\pi} 2\pi I(\theta, 0) \sin\theta \cos\theta \, d\theta$$

$$= 4\pi H. \tag{B–12}$$

From Kirchhoff's law, it follows that

$$j_v = k_v B(v, T) \tag{B–13}$$

with

$$B(v, T) = \frac{2hv^3/c^2}{e^{hv/kT} - 1}.$$

For a constant absorption coefficient, we have

$$J = \frac{j}{k} = \int B(v, T) \, dv = \frac{\sigma}{\pi} T^4. \tag{B–14}$$

Equations (B–11) and (B–14) imply

$$\frac{\sigma}{\pi} T^4 = H(2 + 3\tau). \tag{B–15}$$

The effective temperature T_e is defined to be the temperature of a surface emitting an amount of radiation equal to that from the star. From Equation (B–15) we find

$$T^4 = \frac{T_e^4}{2} (1 + \tfrac{3}{2}\tau). \tag{B–16}$$

Equation (B–16) shows that the effective temperature and actual temperature are equal at $\tau = \tfrac{2}{3}$.

Equation
of Ionization Equilibrium
(Saha Equation)

The equilibrium ratio of the populations of two states separated by energy E is given by the well-known Boltzmann formula

$$\frac{n_1}{n_0} = \frac{g_1}{g_0} e^{-E/kT} \qquad \text{(C–1)}$$

where g_0 and g_1 denote the degeneracy of the lower and upper states, respectively. We consider the specific example of the ionization equilibrium of hydrogen, that is,

$$H \rightleftarrows p + e^- \qquad \text{(C–2)}$$

and assume that the hydrogen atom has only the lowest bound state.

It follows from Equation (C–1) that the ratio of the number density of ionized atoms with free electrons having momenta between p and $p + dp$ to the number density of un-ionized hydrogen atoms is

$$\frac{n_p n_e(p)\, dp}{n_H} = \frac{g_1 g_e}{g_0} e^{-[(X + \frac{1}{2}mv^2)/kT]} \qquad \text{(C–3)}$$

where X is the ionization potential of hydrogen; $g_1 = 2$ is the degeneracy of the proton; $g_0 = 4$ is the degeneracy of the hydrogen ground state; and

$$g_e = 2\,\frac{d^3 p}{h^3} \qquad \text{(C–4)}$$

is the number of free electron states in a unit volume. Using the relation

$$4\pi m^3 v^2\, dv = 4\pi p^2\, dp \qquad \text{(C–5)}$$

and integrating over the Maxwellian velocity distribution of the electrons, we find

$$\frac{n_p n_e}{n_H} = \left(\frac{2\pi m k T}{h^2}\right)^{3/2} \frac{2g_1}{g_0} e^{-X/kT} \tag{C–6}$$

where m is the mass of the electron. It is clear that the above discussion can be generalized to include the effects of additional bound states as well as the ionization of more complicated atoms and the dissociation of molecules. For the case of the ionization of hydrogen and helium, the effect of bound states other than the ground state is small.

APPENDIX D

Gravitational Radiation

General relativity predicts the existence of gravitational waves that propagate with the speed of light. Gravitational waves are polarized and, like electromagnetic waves, cause acceleration that is at right angles to the direction of propagation. A polarized electromagnetic wave will cause electrons to vibrate in the same direction. This direction and the direction of propagation of the wave determine the plane of polarization. On the other hand, the polarization of gravitational waves cannot be characterized by a single plane of polarization but, as in the case of shear waves, requires that two mutually perpendicular planes of polarization be defined. Gravitational radiation also differs from electromagnetic radiation in that quadrupole radiation is the lowest-order radiative process. This circumstance implies that a rotating body will radiate gravitational waves only if it is nonaxially symmetric.

A rigorous discussion of gravitational radiation could, of course, require that we utilize the tensor field equations of general relativity. Here we will only describe some of the properties of gravitational waves by means of a simple and heuristic generalization of Newtonian Theory.[1] The Poisson equation

$$\nabla^2 \Phi = +4\pi G\rho \qquad (D-1)$$

describes the Newtonian relation between the gravitational potential Φ and the density of matter ρ. It can be shown that some components of the tensor field equations of general relativity will satisfy the simplest generalization of Equation (D-1), namely,

$$\nabla^2 \Phi - \frac{1}{c^2} \frac{\partial^2 \Phi}{\partial t^2} = +4\pi G\rho. \qquad (D-2)$$

---

[1] A more rigorous discussion of gravitational radiation is given in Weber, J., 1961, *General Relativity and Gravitational Waves* (New York: Interscience) and Landau, L. and Lifshiftz, E., 1962, *The Classical Theory of Fields* (Oxford: Pergamon Press).

As in classical electromagnetic theory, a solution of the above equation is

$$\Phi(r, t) = -G \int \frac{\rho[r', t - (|r - r'|/c)] \, d^3r'}{|r - r'|}. \tag{D-3}$$

If the size of the source is much smaller than the distance between the source and the observation at position $\mathbf{r}$ and unit direction $\hat{n}$, we can employ the approximation

$$\rho\left(r', t - \frac{|r - r'|}{c}\right) \simeq \rho\left(r', t - \frac{r}{c} + \frac{\hat{n} \cdot \mathbf{r}'}{c}\right)$$

$$\simeq \rho\left(r', t - \frac{r}{c}\right) + \frac{\hat{n} \cdot \mathbf{r}'}{c} \dot{\rho}\left(r', t - \frac{r}{c}\right)$$

$$+ \frac{1}{2}\left(\frac{\hat{n} \cdot \mathbf{r}'}{c}\right)^2 \ddot{\rho}\left(r', t - \frac{r}{c}\right) + \cdots. \tag{D-4}$$

Substituting the above expression for ρ into Equation (D–3), we find

$$\Phi(r, t) = -G\left(\frac{M}{r} + \frac{\hat{n} \cdot \mathbf{P}}{cr} + \text{quadrupole term} + \cdots\right) \tag{D-5}$$

where M is the mass, and P is the momentum. In analogy with the theory of electromagnetic waves, it is plausible to assume that the radiated power has the form

$$-\frac{dE}{dt} \simeq \frac{c}{G} \int \frac{1}{c^2} \dot{\Phi}^2 \, dA \tag{D-6}$$

where the integration is carried out over a sphere at large distances, and the coefficient c/G is chosen to make the dimensions correct. It follows from Equations (D–5) and (D–6) and the constancy of M and $\mathbf{P}$ that quadrupole radiation is the lowest-order radiation term.

For two bodies of equal mass m moving in a circular orbit of diameter D, the quadrupole term in Equation (D–5) is approximately

$$\frac{Gm\omega^2 D^2}{rc^2}. \tag{D-7}$$

It follows from expressions (D–6) and (D–7) that the radiated power is

$$-\frac{dE}{dt} \sim \frac{Gm^2\omega^6 D^4}{c^5}. \tag{D-8}$$

In obtaining (D–7) and (D–8) we have used the expression

$$\dot{\rho} \equiv \frac{d\rho}{dt} = \omega\rho.$$

Some Useful Quantities

PHYSICAL CONSTANTS

Gravitational constant	$G = 6.673 \times 10^{-8}$ dyn cm^2 g^{-2}
Speed of light	$c = 2.9979 \times 10^{10}$ cm sec^{-1}
Planck's constant	$h = 6.625 \times 10^{-27}$ erg sec
	$h/2\pi = \hbar = 1.054 \times 10^{-27}$ erg sec
Boltzmann's constant	$k = 1.3806 \times 10^{-16}$ erg $^\circ$K^{-1}
Blackbody constant	$a = 7.564 \times 10^{-15}$ erg cm^{-3} $^\circ$K^{-4}
Stefan Boltzmann constant	$\sigma = ac/4 = 5.67 \times 10^{-5}$ dyn cm^{-2} $^\circ$K^{-4}
Avogadro's number	$N_0 = 6.02 \times 10^{23}$ mol^{-1}
Electron volt	$1 \text{ eV} = 1.602 \times 10^{-12}$ erg
Electron charge	$e = 4.803 \times 10^{-10}$ esu
Fine structure constant	$\alpha = e^2/\hbar c = 1/137.036$
Electron mass	$m = 9.1095 \times 10^{-28}$ g
Proton mass	$m_p = 1.67 \times 10^{-24}$ g
Thomson cross section	$\sigma_T = 0.665 \times 10^{-24}$ cm^2
Rydberg	$R_y = 13.6058$ eV

ASTRONOMICAL QUANTITIES

Mass of sun	$M\odot = 1.99 \times 10^{33}$ g
Luminosity of sun	$L\odot = 3.90 \times 10^{33}$ erg sec^{-1}
Radius of sun	$R\odot = 6.96 \times 10^{10}$ cm
Effective temperature of sun	$T_e\odot = 5780$ $^\circ$K
Parsec	$1 \text{ pc} = 3.0856 \times 10^{18}$ cm
Light year	$1 \text{ light year} = 9.4605 \times 10^{17}$ cm
Mean earth-sun distance	$1 \text{ AU} = 1.49598 \times 10^{13}$ cm

References

CHAPTER 1

Blaauw, A., Schmidt, M. (eds.) 1965, *Galactic Structure* (Chicago: University of Chicago Press).
Bondi, H. 1960, *Cosmology* (London: Cambridge University Press).
Dalgarno, A., McCray, R. A. 1972, *Ann. Rev. Astron. Ap.*, **10**, 375.
Dupree, A. K., Goldberg, L. 1971, *Ann. Rev. Astron. Ap.*, **8**, 231.
Field, G. B., Somerville, W. B., Dressler, K. 1966, *Ann. Rev. Astron .Ap.*, **4**, 207.
Field, G. B., Goldsmith, D. W., Habing, H. J. 1969, *Ap. J. Lett.*, **155**, L49.
Field, G. B. 1972, *Ann. Rev. Astron. Ap.*, **10**, 227.
Kaplan, S. A., Pikelner, S. B. 1970, *The Interstellar Medium* (Cambridge: University Press).
Kerr, F. J. 1969, *Ann. Rev. Astron. Ap.*, **7**, 39.
Mathews, W. G., O'Dell, C. R. 1969, *Ann. Rev. Astron. Ap.*, **7**, 67.
Middlehurst, B. M., Aller, L. H. (eds.) 1968, *Nebulae and Interstellar Matter* (Chicago: University of Chicago Press).
Mihalas, D., Routly, P. 1968, *Galactic Astronomy* (San Francisco: W. H. Freeman and Company).
Osterbrock, D. E. 1962, *Ap. J.*, **135**, 195.
Purcell, E. M., Field, G. B. 1957, *Ap. J.*, **124**, 542.
Purcell, E. M. 1969, *Ap. J.*, **158**, 433.
Rank, D. M., Townes, C. H., Welch, W. J. 1971, *Science*, **174**, 1083.
Robinson, B. J., McGee, R. X. 1967, *Ann. Rev. Astron. Ap.*, **5**, 183.
Spitzer, L. 1968, *Diffuse Matter in Space* (New York: Interscience Publishers).

CHAPTER 2

Bahcall, J. N., Sears, R. L. 1972, *Ann. Rev. Astron. Ap.*, **10**, 25.
Chiu, H-Y. 1968, *Stellar Physics* (Waltham: Blaisdell Publishing Company).

Chiu, H-Y., Muriel, A. (eds.) 1972, *Stellar Evolution* (Cambridge: M.I.T. Press).

Clayton, D. 1968, *Principles of Stellar Evolution and Nucleosynthesis* (New York: McGraw-Hill, Inc.).

Cox, J. P., Giuli, R. T. 1968, *Principles of Stellar Structure* (New York: Gordon and Breach).

Dicke, R. H. 1969, *Ap. J.*, **155**, 123.

Faulkner, J., Iben, I. 1966, *Ap. J.*, **144**, 995.

Fowler, W. A., Hoyle, F. 1960, *Ann. Phys.*, **10**, 280.

Fricke, K., Kippenhahn, R. 1972, *Ann. Rev. Astron. Ap.*, **10**, 45.

Iben, I. I. 1967, *Ann. Rev. Astron. Ap.*, **5**, 571.

Iben, I. I. 1971, *PASP*, **83**, 697.

Mihalas, D. 1970, *Stellar Atmospheres* (San Francisco: W. H. Freeman and Company).

Paczyński, B. 1971, *Ann. Rev. Astron. Ap.*, **9**, 183.

Parker, E. N. 1969, *Space Sci. Rev.*, **9**, 325.

Schramm, D. N., Wasserburg, G. J. 1971, *Ap. J.*, **162**, 57.

Schwarzshild, M. 1958, *Structure and Evolution of the Stars* (Princeton: Princeton University Press).

Spiegel, E. A. 1971, *Ann. Rev. Astron. Ap.*, **9**, 323.

CHAPTER 3

Chandrasekhar, S., Fermi, E. 1953, *Ap. J.*, **118**, 116.

Hayashi, C., Nakano, T. 1965, *Progress of Theoretical Physics*, **34**, 754.

Hayashi, C. 1966, *Ann. Rev. Astron. Ap.*, **4**, 171.

Heiles, C. 1971, *Ann. Rev. Astron. Ap.*, **9**, 293.

Hunter, C. 1962, *Ap. J.*, **136**, 594.

Larson, R. B., 1969, *M.N.*, **145**, 271.

Spitzer, L. 1968, *Diffuse Matter in Space* (New York: Interscience Publishers).

CHAPTER 4

Baker, N., Kippenhahn, R. 1965, *Ap. J.*, **142**, 868.

Christy, R. F. 1966, *Ap. J.*, **144**, 108.

Christy, R. F. 1966, *Ann. Rev. Astron. Ap.*, **4**, 353.

Cox, J. P., Giuli, R. T. 1968, *Principles of Stellar Structure* (New York: Gordon and Breach).

Harper, R., Rose, W. 1970, *Ap. J.*, **162**, 963.

Iben, I. I. 1971, *PASP*, **83**, 697.

Ledoux, P., Walraven, Th. 1958, *Handbuch der Physik*, **51**, 353.

CHAPTER 5

Chiu, H-Y., Muriel, A. (eds.) 1972, *Stellar Evolution* (Cambridge: M.I.T. Press).

Rose, W. K., Smith, R. L. 1972, *Ap. J.*, **173**, 385.

Schwarzschild, M. 1958, *Structure and Evolution of the Stars* (Princeton: Princeton University Press).
Schwarzschild, M., Härm, R. 1965, *Ap. J.*, **142**, 855.
Smith, R. L., Rose, W. K. 1972, *Ap. J.*, **176**, 395.
Weigert, A. 1966, *Zs. f. Ap.*, **64**, 395.

CHAPTER 6

Giacconi, R., Gursky, H., Van Speybroeck 1968, *Ann. Rev. Astron. Ap.*, **6,** 373.
Harper, R., Rose, W. K. 1970, *Ap. J.*, **162**, 963.
Hazelhurst, J. 1962, *Adv. Ast. Ap.*, **1,** 1.
Kraft, R. P. 1964, *Ap. J.*, **139**, 457.
McLaughlin, D. 1960, *Stellar Atmospheres*, J. Greenstein (ed.) (Chicago: University of Chicago Press).
Morrison, P. 1967, *Ann. Rev. Astron. Ap.*, **5,** 325.
Pottasch, S. 1959, *Ann. d'ap.*, **22**, 297.
Rose, W. K. 1968, *Ap. J.*, **152**, 245.
Rose, W. K., Smith, R. L. 1972, *Ap. J.*, **172**, 699.
Schatzman, E. 1965, *Stellar Structure*, J. Allen and D. McLaughlin (eds.) (Chicago: University of Chicago Press).
Starrfield, S., Truran, J. W., Sparks, W. M., Kutter, G. S. 1972, *Ap. J.*, **176**, 169.
Warner, B., Robinson, E. L. 1972, *Nature*, **239**, 2.
Zel'dovich, Ya. B. and Novikov, I. 1971, *Relativistic Astrophysics*, vol. 1 (Chicago: University of Chicago Press).

CHAPTER 7

Gurzadyan, G. A. 1969, *Planetary Nebulae* (New York: Gordon and Breach).
Harmon, R. J., Seaton, M. J. 1964, *Ap. J.*, **140**, 824.
O'Dell, C. R. 1963, *Ap. J.*, **138**, 67.
O'Dell, C. R., Osterbrock, D. E. (eds.) 1968, *Planetary Nebulae* (Dordrecht, Holland: Reidel).
Rose, W. K. 1967, *Ap. J.*, **150**, 193.
Rose, W. K., Smith, R. L. 1970, *Ap. J.*, **159**, 903.
Salpeter, E. E. 1971, *Ann. Rev. Astron. Ap.*, **9,** 127.
Seaton, M. J. 1958, *Rev. Mod. Phy.*, **30**, 979.
Seaton, M. J. 1960, *Rep. Prog. Phys.*, **23**, 313.
Seaton, M. J. 1966, *MNRAS*, **132**, 113.
Smith, R. L., Rose, W. K. 1972, *Ap. J.*, **176**, 395.

CHAPTER 8

Chandrasekhar, S. 1957, *Stellar Structure* (New York: Dover Publications, Inc.)

Mestel, L., Ruderman, M. A. 1967, *MNRAS*, **136**, 27.
Ostriker, J. P. 1971, *Ann. Rev. Astron. Ap.*, **9**, 353.
Salpeter, E. E. 1961, *Ap. J.*, **134**, 669.
Van Horn, H. M. 1968, *Ap. J.*, **151**, 227.
Weidemann, V. 1968, *Ann. Rev. Astron. Ap.*, **6**, 351.

CHAPTER 9

Arnett, W. D. 1969, *Ap. and Space Sci.*, **5**, 180.
Barkat, Z., Wheeler, J. C., Buchler, J. R. 1972, *Ap. J.*, **171**, 651.
Bruenn, S. W. 1972, *Ap. J.*, **177**, 460.
Burbidge, E. M., Burbidge, G. R., Fowler, W. A., Hoyle, F. 1957, *Rev. Mod. Phys.*, **28**, 547.
Clayton, D. 1968, *Principles of Stellar Evolution and Nucleosynthesis* (New York: McGraw-Hill, Inc.).
Colgate, S. A. 1971, *Ap. J.*, **163**, 221.
Felten, J., Morrison, P. 1966, *Ap. J.*, **146**, 686.
Ginzburg, V. L. 1969, *Elementary Processes for Cosmic Ray Astrophysics* (New York: Gordon and Breach).
Morrison, P., Sartori, L. 1969, *Ap. J.*, **158**, 541.
Paczyński, B. 1970, *Acta. Astr.*, **20**, 47.
Rakavy, G., Shaviv, G., Zinamon, Z. 1967, *Ap. J.*, **150**, 131.
Rose, W. K. 1969, *Ap. J.*, **155**, 491.
Shklovsky, I. S. 1968, *Supernovae* (New York: John Wiley & Sons, Inc.).
Trimble, V., Rees, M. J. 1970, *Ap. Lett.*, **5**, 93.
Wilson, J. 1971, *Ap. J.*, **163**, 209.
Woltjer, L. 1972, *Ann. Rev. Astron. Ap.*, **10**, 129.

CHAPTER 10

Goldreich, P., Julian, W. H. 1969, *Ap. J.*, **157**, 869.
Gunn, J. E., Ostriker, J. P. 1969, *Nature*, **221**, 454.
Gunn, J. E., Ostriker, J. P. 1971, *Ap. J.*, **165**, 523.
Hewish, A. 1970, *Ann. Rev. Astron. Ap.*, **8**, 265.
Komesaroff, M. M. 1970, *Nature*, **225**, 612.
Pacini, F. 1968, *Nature*, **219**, 145.
Pacini, F., Rees, M. J. 1970, *Nature*, **226**, 622.
Ruderman, M. 1972, *Ann. Rev. Astron. Ap.*, **10**, 427.
Woltjer, L., Setti, G. 1970, *Ap. J.*, **159**, L87.

CHAPTER 11

Bahcall, J. N., Wolf, R. A., *Phys. Rev.*, **140**, B1445.
Cameron, A. G. W. 1970, *Ann. Rev. Astron. Ap.*, **8**, 179.

Harrison, B. K., Thorne, K. S., Wakano, M., Wheeler, J. A. 1965, *Gravitational Theory and Gravitational Collapse* (Chicago: University of Chicago Press).

Tsuruta, S., Cameron, A. G. W. 1966, *Canadian J. Phys.*, **44**, 1863.

Wang, C. G., Rose, W. K., Schlenker, S. L. 1970, *Ap. J.*, **160**, L17.

Zel'dovich, Ya. B., Novikov, I. 1971, *Relativistic Astrophysics*, vol. 1 (Chicago: University of Chicago Press).

CHAPTER 12

Burbidge, G., Burbidge, M. 1967, *Quasi-Stellar Objects* (San Francisco: W. H. Freeman and Company).

Evans, S. E. (ed.) 1972, *External Galaxies and Quasi-Stellar Objects* (Dordrecht, Holland: Reidel).

Hodge, P. 1966, *Galaxies and Cosmology* (New York: McGraw-Hill, Inc.).

Kellerman, K. I., Pauling-Toth, I. I. K. 1968, *Ann. Rev. Astron. Ap.*, **6**, 417.

Lin, C. C., Shu, F. H. 1964, *Ap. J.*, **140**, 646.

Morrison, P., Sartori, L., *Ap. J.*, **152**, L139.

Pacholczyk, A. G. 1970, *Radio Astrophysics* (San Francisco: W. H. Freeman and Company).

Rees, M. J. 1967, *MNRAS*, **135**, 345.

Schmidt, M. 1969, *Ann. Rev. Astron. Ap.*, **7**, 527.

Swihart, T. L. 1968, *Astrophysics and Stellar Astronomy* (New York: John Wiley & Sons, Inc.).

CHAPTER 13

Dicke, R. H., Peebles, P. J. E., Roll, P. G., Wilkinson, D. T. 1965, *Ap. J.*, **142,** 414.

Misner, C. W., Thorne, K. S., Wheeler, J. A. 1972, *Gravitation* (San Francisco: W. H. Freeman and Company).

Peebles, P. J. E. 1969, *Am. J. Phys.*, **37,** 410.

Peebles, P. J. E. 1971, *Physical Cosmology* (Princeton: Princeton University Press).

Penzias, A. A., Wilson, R. W. 1965, *Ap. J.*, **142,** 419.

Robertson, H. P., Noonan, T. W. 1968, *Relativity and Cosmology* (Philadelphia: W. B. Saunders Company).

Ryle, M. 1968, *Ann. Rev. Astron. Ap.*, **6**, 249.

Sandage, A. R. 1961, *Ap. J.*, **133,** 355.

Sciama, D. W. 1971, *Modern Cosmology* (Cambridge: Cambridge University Press).

Thaddeus, P. 1972, *Ann. Rev. Astron. Ap.*, **10,** 305.

Weinberg, S. 1972, *Gravitation and Cosmology* (New York: John Wiley & Sons, Inc.).

INDEX